Food Safety and Natural Toxins

Food Safety and Natural Toxins

Special Issue Editors

Mary Fletcher
Gabriele Netzel

MDPI • Basel • Beijing • Wuhan • Barcelona • Belgrade • Manchester • Tokyo • Cluj • Tianjin

Special Issue Editors

Mary Fletcher
University of Queensland
Australia

Gabriele Netzel
University of Queensland
Australia

Editorial Office
MDPI
St. Alban-Anlage 66
4052 Basel, Switzerland

This is a reprint of articles from the Special Issue published online in the open access journal *Toxins* (ISSN 2072-6651) (available at: https://www.mdpi.com/journal/toxins/special_issues/Natural_Toxins_Food_Safety).

For citation purposes, cite each article independently as indicated on the article page online and as indicated below:

LastName, A.A.; LastName, B.B.; LastName, C.C. Article Title. *Journal Name* **Year**, *Article Number*, Page Range.

ISBN 978-3-03936-445-9 (Hbk)
ISBN 978-3-03936-446-6 (PDF)

Contents

About the Special Issue Editors vii

Mary T. Fletcher and Gabriele Netzel
Food Safety and Natural Toxins
Reprinted from: *Toxins* **2020**, *12*, 236, doi:10.3390/toxins12040236 1

Gabriele Netzel, Eddie T. T. Tan, Mukan Yin, Cindy Giles, Ken W. L. Yong, Rafat Al Jassim and Mary T. Fletcher
Bioaccumulation and Distribution of Indospicine and Its Foregut Metabolites in Camels Fed *Indigofera spicata*
Reprinted from: *Toxins* **2019**, *11*, 169, doi:10.3390/toxins11030169 3

Saira Sultan, Cindy Giles, Gabriele Netzel, Simone A. Osborne, Michael E. Netzel and Mary T. Fletcher
Release of Indospicine from Contaminated Camel Meat following Cooking and Simulated Gastrointestinal Digestion: Implications for Human Consumption
Reprinted from: *Toxins* **2018**, *10*, 356, doi:10.3390/toxins10090356 11

Agha Waqar Yunus, Nida Imtiaz, Haider Khan, Mohammed Nawaz Mohammed Ibrahim and Yusuf Zafar
Aflatoxin Contamination of Milk Marketed in Pakistan: A Longitudinal Study
Reprinted from: *Toxins* **2019**, *11*, 110, doi:10.3390/toxins11020110 23

AleksandarŽ. Kostić, Danijel D. Milinčić, Tanja S. Petrović, Vesna S. Krnjaja, Sladjana P. Stanojević, Miroljub B. Barać,Živoslav Lj. Tešić and Mirjana B. Pešić
Mycotoxins and Mycotoxin Producing Fungi in Pollen: Review
Reprinted from: *Toxins* **2019**, *11*, 64, doi:10.3390/toxins11020064 33

Kristina Mastanjević, Jasmina Lukinac, Marko Jukić, Bojan Šarkanj, Vinko Krstanović and Krešimir Mastanjević
Multi-(myco)toxins in Malting and Brewing By-Products
Reprinted from: *Toxins* **2019**, *11*, 30, doi:10.3390/toxins11010030 53

Cheryl M. Armstrong, Leah E. Ruth, Joseph A. Capobianco, Terence P. Strobaugh Jr., Fernando M. Rubio and Andrew G. Gehring
Detection of Shiga Toxin 2 Produced by *Escherichia coli* in Foods Using a Novel AlphaLISA
Reprinted from: *Toxins* **2018**, *10*, 422, doi:10.3390/toxins10110422 69

Mélanie Roué, Hélène Taiana Darius, André Ung, Jérôme Viallon, Manoella Sibat, Philipp Hess, Zouher Amzil and Mireille Chinain
Tissue Distribution and Elimination of Ciguatoxins in *Tridacna maxima* (*Tridacnidae*, Bivalvia) Fed *Gambierdiscus polynesiensis*
Reprinted from: *Toxins* **2018**, *10*, 189, doi:10.3390/toxins10050189 83

About the Special Issue Editors

Mary Fletcher (Associate Professor) received her PhD degree in Natural Product Organic Chemistry from The University of Queensland (Australia, 1986), where she currently leads the Natural Toxins Research Group within the Queensland Alliance for Agriculture and Food Innovation. Her research interests are in the identification and analysis of natural toxins in a range of plants, fungi, and agricultural products, with a focus on potential impacts of these toxins on both human and animal health and risks posed to livestock production, food safety, and market access.

Gabriele Netzel received her doctoral degree in Nutritional Sciences and Home Economics from the Justus-Liebig University Giessen (Germany, 2011). She was awarded a Postdoctoral Fellowship from the Queensland Government (Advance Queensland Research Fellowship), where she investigated the detrimental effects of the natural toxin indospicine on cattle. She has recently joined the ARC (Australian Research Council) Industrial Transformation Training Centre for Uniquely Australian Foods within the Queensland Alliance for Agriculture and Food Innovation at The University of Queensland Research project. Her research now focuses on the assessment of the nutritional quality of Australian native fruits and vegetables.

toxins

MDPI

Editorial

Food Safety and Natural Toxins

Mary T. Fletcher * and Gabriele Netzel

Queensland Alliance for Agriculture and Food Innovation (QAAFI), The University of Queensland, Health and Food Sciences Precinct, Coopers Plains, QLD 4108, Australia; g.netzel@uq.edu.au

* Correspondence: mary.fletcher@uq.edu.au; Tel.: +61-7-3443-2479

Received: 11 February 2020; Accepted: 31 March 2020; Published: 8 April 2020

Natural toxins are poisonous secondary metabolites produced by living organisms, which are typically not harmful to the organisms themselves but can impact on human or animal health when consumed [1,2]. Common sources of such toxins include poisonous plants, fungi, algae, bacteria and marine biotoxins, and the diversity of these biological systems presents challenges to analytical chemists and wide-ranging food safety implications. The propensity for such toxins to be present in both animal feed and human food has led to the introduction of regulations for a small number of the most potent natural toxins, particularly mycotoxins [3]. Implementation of these regulations necessitates the establishment of high-throughput analytical chemistry methods [4], such as LC-MS and GC-MS, with increasingly lower limits of detection, and the production of isotopically labelled analogues as internal standards to increase the reliability of analysis in complex matrices. Thus, this Special Issue focused on the analysis of natural toxins, their incidence from source organisms to food and feed commodities, and implications for food safety.

The impacts of natural toxins present in pasture plants are not limited to grazing livestock but can also be carried through the food chain, with some toxins, such as indospicine, being demonstrated to accumulate in the tissues of grazing animals and cause secondary poisoning in animals consuming meat from these livestock [5,6]. Interestingly, camel feeding trials have demonstrated that indospicine is still detectable by LC-MS/MS analysis 100 days after cessation of feeding *Indigofera* plants, whereas ruminal metabolites of this toxin have a considerably faster clearance rate [5]. The toxicity of indospicine for humans is uncertain but simple cooking of indospicine-contaminated meat does not degrade this natural toxin, and in vitro studies under human gastrointestinal conditions further demonstrated its stability and release into liquid digesta indicating ready bio-accessibility for absorption in the small intestine [6].

Mycotoxins are natural toxins produced by fungi in a range of food and feed commodities, with potent carcinogenic aflatoxins (particularly aflatoxin B_1 or AFB1) being considered of significant risk to human health. Once ingested by animals, AFB1 can be carried into milk as the toxic metabolite aflatoxin M_1, with the presence of AFM1 in milk being of particular concern as young children represent a particularly susceptible age group [7]. AFM1 analysis studies have assessed the risk of AFM1 in both liquid and powdered milk marketed in Pakistan, with processed milk shown to have lower AFM1 levels than raw milk [7]. Mycotoxin risk is not only limited to the consumption of cereal grains, but can also occur in pollen as used in food supplements [8], and malting and brewing by-products as used in animal feed [9], and reviews of these products further highlight the potential mycotoxin risks in these novel food and feed products.

Natural toxins produced by bacteria are also of concern. Shiga toxin-producing *Escherichia coli* is a frequent cause of food poisoning with enormous human health and economic impacts, and rapid and accurate identification is imperative for protection of human health. Detection of the toxin by a newly developed amplified luminescent proximity homogenous assay-linked immunosorbent assay (AlphaLISA) allows for an automated and much more rapid assay of the toxin with increased sensitivity and dynamic range compared to the industry-standard ELISA test [10].

Another class of natural toxins are ciguatoxins (CTXs), marine biotoxins produced by dinoflagellates of the genus *Gambierdiscus*, with the foodborne disease ciguatera caused by the consumption of contaminated seafood. Analytical studies have examined both CTXs' tissue distribution in giant clams (*Tridacna maxima*) exposed to toxic cells of *Gambierdiscus polynesiensis*, and also detoxification rates after the transfer of clams into clean water [11]. Viscera, flesh, and mantle contained 65%, 25%, and 10% of the toxin burden respectively, with all tissues reaching levels above safety limits, and most concerningly, no toxin elimination was seen within a six day detoxification period.

This Special Issue then contains original contributions that advance our knowledge of the food safety implications of natural toxins. The breadth of the manuscripts demonstrated the diverse sources of natural toxins that can affect food safety, from plant toxins and their accumulation in the meat of grazing livestock, mycotoxins and their presence in grains, brewing by-products, milk and pollen supplements, ciguatoxins in seafood, and Shiga toxins associated with *Escherichia coli* contamination of foods.

Acknowledgments: The guest editors express their appreciation to all the authors who contributed to this Special Issue. They are also grateful for the rigorous and selfless evaluation of the submitted manuscripts by expert peer reviewers, without which this special issue would not be possible. The valuable editorial support provided by the MDPI management team and staff are also greatly appreciated.

Conflicts of Interest: The authors declare no conflict of interest.

References

1. WHO. Natural Toxins in food (Factsheet). 2018. Available online: https://www.who.int/news-room/fact-sheets/detail/natural-toxins-in-food (accessed on 10 February 2020).
2. Dolan, L.C.; Matulka, R.A.; Burdock, G.A. Naturally occurring food toxins. *Toxins* **2010**, *2*, 2289–2332. [CrossRef] [PubMed]
3. Food and Agriculture Organization of the United Nations (FAO); The World Health Organization (WHO). Codex Allimentarius International Food Standards. General standard for contaminants and toxins in food and feed. (Last amended 2019). *CODEX STAN*, 193–1995. Available online: http://www.fao.org/fao-who-codexalimentarius/codex-texts/list-standards/en/ (accessed on 10 February 2020).
4. Rodríguez, I.; González, J.M.; Botana, A.M.; Sainz, M.J.; Vieytes, M.R.; Alfonso, A.; Botana, L.M. Analysis of natural toxins by liquid chromatography. In *Liquid Chromatography: Applications*, 2nd ed.; Fanali, S., Haddad, P.R., Poole, C.F., Riekkola, M.-L., Eds.; Elsevier: Amsterdam, The Netherlands, 2017; pp. 479–514.
5. Netzel, G.; Tan, E.T.T.; Yin, M.; Giles, C.; Yong, K.W.L.; Al Jassim, R.; Fletcher, M.T. Bioaccumulation and distribution of indospicine and its foregut metabolites in camels fed *Indigofera spicata*. *Toxins* **2019**, *11*, 169. [CrossRef] [PubMed]
6. Sultan, S.; Giles, C.; Netzel, G.; Osborne, S.A.; Netzel, M.E.; Fletcher, M.T. Release of indospicine from contaminated camel meat following cooking and simulated gastrointestinal digestion: Implications for human consumption. *Toxins* **2018**, *10*, 356. [CrossRef] [PubMed]
7. Yunus, A.W.; Imtiaz, N.; Khan, H.; Ibrahim, M.N.M.; Zafar, Y. Aflatoxin contamination of milk marketed in Pakistan: A longitudinal study. *Toxins* **2019**, *11*, 110. [CrossRef] [PubMed]
8. Kostić, A.Ž.; Milinčić, D.D.; Petrović, T.S.; Krnjaja, V.S.; Stanojević, S.P.; Barać, M.B.; Tešić, Ž.L.; Pešić, M.B. Mycotoxins and mycotoxin producing fungi in pollen: Review. *Toxins* **2019**, *11*, 64. [CrossRef] [PubMed]
9. Mastanjević, K.; Lukinac, J.; Jukić, M.; Šarkanj, B.; Krstanović, V.; Mastanjević, K. Multi-(myco)toxins in malting and brewing by-products. *Toxins* **2019**, *11*, 30. [CrossRef] [PubMed]
10. Armstrong, C.M.; Ruth, L.E.; Capobianco, J.A.; Strobaugh, T.P.; Rubio, F.M.; Gehring, A.G. Detection of Shiga Toxin 2 produced by *Escherichia coli* in foods using a novel AlphaLISA. *Toxins* **2018**, *10*, 422. [CrossRef] [PubMed]
11. Roué, M.; Darius, H.T.; Ung, A.; Viallon, J.; Sibat, M.; Hess, P.; Amzil, Z.; Chinain, M. Tissue distribution and elimination of ciguatoxins in *Tridacna maxima* (Tridacnidae, Bivalvia) fed *Gambierdiscus polynesiensis*. *Toxins* **2018**, *10*, 189. [CrossRef] [PubMed]

toxins

MDPI

Article

Bioaccumulation and Distribution of Indospicine and Its Foregut Metabolites in Camels Fed *Indigofera spicata*

Gabriele Netzel [1], Eddie T. T. Tan [1,2], Mukan Yin [1], Cindy Giles [3], Ken W. L. Yong [3], Rafat Al Jassim [1] and Mary T. Fletcher [1,*]

1. Queensland Alliance for Agriculture and Food Innovation (QAAFI), The University of Queensland, Health and Food Sciences Precinct, Coopers Plains, QLD 4108, Australia; g.netzel@uq.edu.au (G.N.); eddietan@ns.uitm.edu.my (E.T.T.T.); mukan.yin@uq.net.au (M.Y.); r.aljassim@uq.edu.au (R.A.J.)
2. Alliance of Research and Innovation for Food (ARIF), Faculty of Applied Sciences, Universiti Teknologi MARA, Cawangan Negeri Sembilan, Kuala Pilah Campus, Negeri Sembilan 72000, Malaysia
3. Department of Agriculture and Fisheries, Health and Food Sciences Precinct, Coopers Plains, QLD 4108, Australia; cindy.giles@daf.qld.gov.au (C.G.); ken.yong@daf.qld.gov.au (K.W.L.Y.)

* Correspondence: mary.fletcher@uq.edu.au; Tel.: +61-7-3443-2479

Received: 20 February 2019; Accepted: 8 March 2019; Published: 19 March 2019

Abstract: In vitro experiments have demonstrated that camel foregut-fluid has the capacity to metabolize indospicine, a natural toxin which causes hepatotoxicosis, but such metabolism is in competition with absorption and outflow of indospicine from the different segments of the digestive system. Six young camels were fed *Indigofera spicata* (337 μg indospicine/kg BW/day) for 32 days, at which time three camels were euthanized. The remaining camels were monitored for a further 100 days after cessation of this indospicine diet. In a retrospective investigation, relative levels of indospicine foregut-metabolism products were examined by UHPLC-MS/MS in plasma, collected during both accumulation and depletion stages of this experiment. The metabolite 2-aminopimelamic acid could be detected at low levels in almost all plasma samples, whereas 2-aminopimelic acid could not be detected. In the euthanized camels, 2-aminopimelamic acid could be found in all tissues except muscle, whereas 2-aminopimelic acid was only found in the kidney, pancreas, and liver tissues. The clearance rate for these metabolites was considerably greater than for indospicine, which was still present in plasma of the remaining camels 100 days after cessation of *Indigofera* consumption.

Keywords: indospicine; 2-aminopimelamic acid; 2-aminopimelic acid; in vivo; foregut metabolites; camel; food safety

Key Contribution: The indospicine metabolites 2-aminopimelamic acid and 2-aminopimelic acid are formed during foregut microbial metabolism and can be detected in tissues of camels fed *Indigofera spicata*. These metabolites are not as persistent as indospicine in animal tissues after cessation of feeding, and could be used as markers for recent *Indigofera* consumption.

1. Introduction

There are more than 60 *Indigofera* species distributed throughout the arid and semiarid regions of Australia [1–4]. *Indigofera* spp. are leguminous shrubs and herbs which are high in protein, as well as highly palatable for animals. These plants are considered a nutritious animal fodder, however, some species contain indospicine, a non-proteinogenic arginine analogue which causes hepatotoxicosis in sheep, cows, rabbits, and dogs [5–9]. The introduced species, *Indigofera spicata*, has been found to contain high levels of this amino acid. Since there is no known mammalian enzyme which is capable of degrading indospicine [10], it can be toxic to simple-stomached animals. However, ruminants seem to be less susceptible to certain toxins due to the ability of rumen microflora to degrade some of those foodborne toxins. Feral camels in Australia seem to cope better in detoxification of plant toxins from their fodder, compared to domesticated animals [11].

Indospicine is a water-soluble free amino acid and rumen degradation processes compete with presumed rapid passage through the fermentation compartments of the digestive tract. This amino acid has been detected in meat, which indicates that at least a portion of indospicine can by-pass the fermentation processes of the foregut and be absorbed as is [12]. A number of dogs have died of secondary hepatotoxicosis after consuming indospicine-contaminated horse meat [7] and, more recently, camel meat [5], and this has raised food safety concerns. In a recent study, we have shown that indospicine accumulated in camel meat during a feeding trial in which six camels were fed a diet containing *Indigofera spicata* for 32 days, and that indospicine can be detected in plasma as long as three months after removing *Indigofera* from the diet [13].

We also reported previously that microflora of both the bovine rumen and camel foregut fluids have the ability to degrade indospicine in vitro within an incubation period of 48 h [14]. However, the in vitro degradability of indospicine is indicative of the potential degradability, and not the actual degradability, that occurs in the animal system. Factors including the microbial community, residence time of the solid fraction of digesta, and outflow rate of the fluid phase all play an important role. Camels are known to retain low quality fibre diets longer in the foregut compared with ruminant animals. Retention time is always shorter when the diet is of higher quality, which should be the case with lush early season pasture containing *Indigofera spicata* at the start of the wet season. Shift to such diet increases the outflow rate and allows more indospicine to enter the intestines where it then gets absorbed. Indospicine has been shown to be chemically stable and resistant to both acidic and base conditions [15,16]. Since the camel foregut fluid is only mildly acidic, it is most likely that rumen bacteria are responsible for the observed metabolism of indospicine (**1**) into its degradation product 2-aminopimelamic acid (**2**) and, further, to 2-aminopimelic acid (**3**) (Figure 1) [14,17].

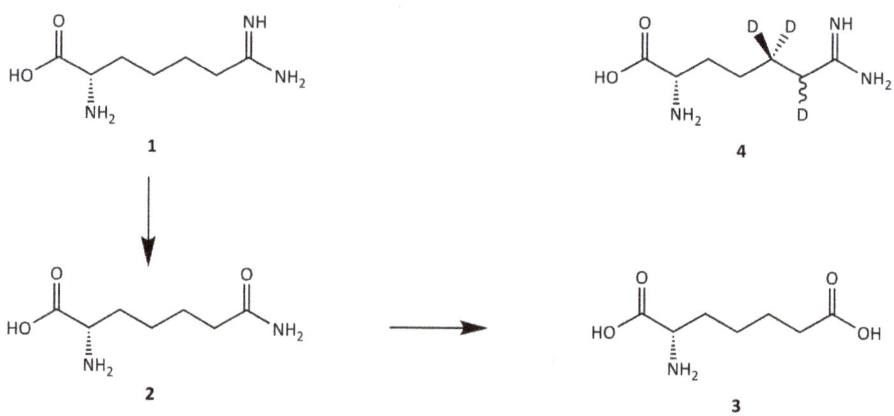

Figure 1. Chemical structures of indospicine (**1**) and its metabolites 2-aminopimelamic acid (**2**) and 2-aminopimelic acid (**3**), together with D_3-L-indospicine (**4**) which is used as an internal standard in LC-MS/MS analysis.

Although we could show previously that indospicine accumulated as a free amino acid in various animal tissues in vivo [13], it has also been demonstrated that indospicine can be metabolized in vitro by foregut microbiota [14]. These two processes of removal (outflow and absorption) and metabolism could be considered to operate in competition, and there is nothing known about the extent of in vivo metabolism of indospicine and whether the metabolites are also transported and accumulated in tissues. Hence, in the present study, we investigated the bioaccumulation and distribution, as well as the excretion, of the indospicine foregut metabolites, 2-aminopimelamic acid and 2-aminopimelic acid, in camels fed *Indigofera* plant material for 32 days.

2. Results and Discussion

2.1. Indospicine and Foregut Metabolites in Tissue Samples

It has previously been established that indospicine accumulates in muscle and other tissues of cattle [18] and camels [13] fed *Indigofera* plant material, however nothing is known about the fate of the indospicine metabolites, 2-aminopimelamic acid and 2-aminopimelic acid. In this study we have measured both indospicine and the two metabolite concentrations in tissues acquired during the previous camel feeding trial, where six young camels (camels 1–6) were fed *Indigofera spicata* for 32 days until indospicine levels in plasma plateaued. At this point, three animals (camels 1–3) were euthanized and the remaining camels (camels 4–6) fed an *Indigofera*-free diet for a further 100 days whilst monitoring decline in indospicine plasma levels [13]. The inclusion rate of *Indigofera spicata* was designed to provide 337 μg of indospicine per kg BW per day.

In accord with the previous analysis, the highest concentration of indospicine in necropsied camels (camels 1–3) was found, in this study, in the pancreas (5.06 ± 0.79 mg/kg FW), followed by the liver (3.57 ± 1.17 mg/kg FW), heart (2.32 ± 0.48 mg/kg FW), kidney (1.48 ± 0.11 mg/kg FW), muscle (1.24 ± 0.26 mg/kg FW), and spleen (0.96 ± 0.34 mg/kg FW) (Figure 2). However, if we look at the deamino metabolites, the highest concentration of the intermediate metabolite 2-aminopimelamic acid was found in the kidney (0.96 ± 0.12 mg/kg FW), followed by the pancreas (0.36 ± 0.11 mg/kg FW), liver (0.27 ± 0.06 mg/kg FW), spleen (0.20 ± 0.08 mg/kg FW), and heart (0.15 ± 0.02 mg/kg FW). Neither 2-aminopimelamic acid nor 2-aminopimelic acid could be detected in the muscle tissue. Only low concentrations of the second metabolite 2-aminopimelic acid could be found in kidney (0.34 ± 0.10 mg/kg FW), pancreas (0.11 ± 0.03 mg/kg FW), and liver (0.01 ± 0.02 mg/kg FW).

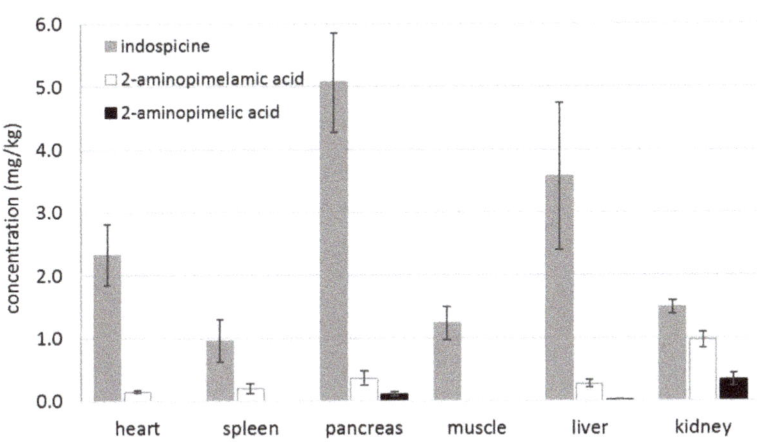

Figure 2. Mean concentrations (mg/kg FW) of indospicine, 2-aminopimelamic acid, and 2-aminopimelic acid for camels 1–3 necropsy tissues (heart, spleen, pancreas, muscle, liver and kidney) after 32-day feeding period.

The kidney, spleen, heart, liver, and pancreas are all organs where arginine metabolism plays an important role in metabolic pathways. Indospicine has, for example, been shown to be a competitive inhibitor of hepatic arginase [19], but arginase does not occur only in the liver. Arginase has two isoforms: arginase I, a cytoplasmic enzyme, which is highly expressed in the liver, and arginase II, a mitochondrial enzyme more widely distributed in extrahepatic tissues, which is expressed primarily in the kidney [20]. Arginase is a key enzyme of the urea cycle and is also present in other tissues such as the spleen, heart, kidney, and pancreas [20–22]. Arginase activity is reportedly blocked through the competitive binding of indospicine [19], which is consistent with the observed elevated presence of indospicine in these same tissues (Figure 2). It is worth noting that 2-aminopimelamic acid is an analogue of citrulline, in the same manner that indospicine is an analogue of arginine. We can hypothesize then that the high concentration of 2-aminopimelamic acid found in the kidney, spleen, heart, liver, and pancreas could likewise be due to interference in citrulline metabolism within the urea cycle, potentially blocking the enzyme argininosuccinate synthetase which utilises citrulline in the synthesis of argininosuccinic acid [22]. The kidney is noted to have a significant capacity for citrulline metabolism [22], and the highest level of accumulation of the citrulline analogue 2-aminopimelamic acid may reflect interference in this metabolism. Such interference could then be consistent with the dramatically elevated citrulline levels reported in indospicine treated rats [23].

2.2. *Indospicine and Foregut Metabolites in Plasma*

The concentration of indospicine [13], as well as that of 2-aminopimelamic acid, was observed to rapidly increase in plasma of camels 1–6 in the first 13–20 days after commencement of the feeding trial (Figure 3). The indospicine concentration reached a plateau phase before it decreased slowly after ceasing the *Indigofera* uptake. The concentration of the 2-aminopimelamic acid similarly rose in the first 13 days, plateauing for about two weeks, before it slightly decreased and rose to a second maximum at 32 days, the point at which the *Indigofera* intake was stopped. In the depletion phase, when the remaining camels (camels 4–6) were fed an *Indigofera*-free diet, the 2-aminopimelamic acid concentration followed the indospicine concentration in a slow decrease until no 2-aminopimelamic acid was detectable at 62 days (30 days after cessation of *Indigofera* consumption).

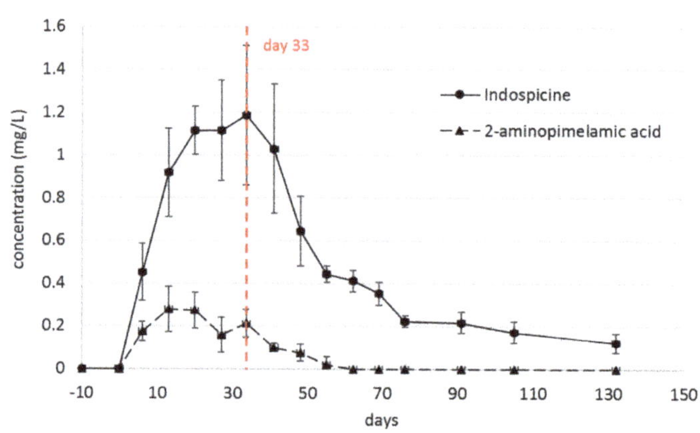

Figure 3. Comparison of indospicine and 2-aminopimelic acid concentrations in plasma (mean ± SD) during the first 32 days (n = 6) of the treatment and after cessation of *Indigofera spicata* feeding (n = 3). Camels 1–3 were autopsied at day 33.

The plasma concentrations of the 2-aminopimelamic acid were much lower than those of indospicine. As already described, there was a considerable variation in the indospicine plasma levels between the individual camels [13], but the individual indospicine plasma curves all followed a similar pattern. By comparison, the pattern of the 2-aminopimelamic acid plasma content proved to be quite variable between the camels (Figure 4). All camels showed a rapid increase within the first 13 or 20 days, to levels between 0.19–0.44 mg/L, with considerable variation between individuals. During the depletion phase, the plasma levels of the three camels (camels 4–6) followed a steady decrease until day 55 or 60, with 2-aminopimelamic acid being eliminated from the system much quicker than indospicine, which is still detected in plasma at 130 days—100 days after cessation of *Indigofera* intake (Figure 3).

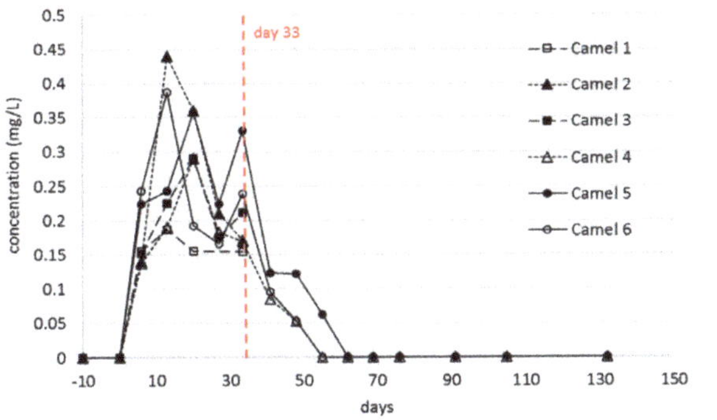

Figure 4. Individual 2-aminopimelamic acid concentrations of plasma during the first 32 days ($n = 6$) of the treatment and after cessation of *Indigofera spicata* feeding ($n = 3$). Camels 1–3 were autopsied at day 33.

The rapid increase in plasma levels of the 2-aminopimelamic acid during the feeding phase followed the rapid increase of indospicine, which can be metabolized in the foregut to 2-aminopimelamic acid and, further, to 2-aminopimelic acid. This second metabolite 2-aminopimelic acid was not detectable in any of the plasma samples. Since no 'fresh' indospicine was uptaken after cessation of the *Indigofera* feeding, it is apparent that the metabolites were eliminated much faster from the camel's system, compared to the indospicine. The lengthier persistence of indospicine residues is attributed to the slow release of this arginine analogue from organ tissue, such as the pancreas, where it was accumulated.

3. Conclusions

These results demonstrate that foregut metabolism of indospicine does occur in vivo, and that the two metabolites, 2-aminopimelamic acid and 2-aminopimelic acid, are both absorbed and bioaccumulate in a range of tissues. After consumption of *Indigofera* plants, the intermediate metabolite 2-aminopimelamic acid was present in all camel tissues except muscle, with highest levels measured in the kidney, pancreas, and liver. Lower levels of 2-aminopimelic acid were also found in these same three tissues. Cytotoxicity studies conducted by Sultan et al. [24] demonstrated that the second metabolite, 2-aminopimelic acid, is less toxic than indospicine. However, nothing to date is known about the toxicity of its precursor, 2-aminopimelamic acid, so there is a possibility that this intermediate metabolite, which is an analogue of citrulline, also contributes to the effects of *Indigofera* poisoning. The clearance of this metabolite from plasma occurs faster compared to indospicine (after cessation of *Indigofera* intake), therefore the presence of 2-aminopimelamic acid in plasma could potentially be used as an indicator of whether the indospicine contamination is recent, within the past month, or longer. The actual route of excretion of indospicine (or its metabolites) has not been investigated, and it is recommended that further trials be conducted to determine levels of indospicine and its metabolites in both milk and urine of animals consuming *Indigofera*.

4. Materials and Methods

4.1. Standards and Reagents

Indospicine (**1**) and 2-aminopimelamic acid (**2**) (>99% pure), both external standards, as well as D_3-L-indospicine (**4**) (>99% pure) as internal standard, were synthesized and provided by Prof. James De Voss and Dr. Robert Lang, The University of Queensland [16,25]. Another external standard, 2-aminopimelic acid (**3**) (>99% pure), and heptafluorobutyric acid (HFBA), ion chromatography grade, were purchased from Sigma Aldrich (Castle Hill, NSW, Australia). External (0.005–2 mg/L) and internal (1 mg/L) standard solutions were prepared in 0.1% HFBA in Milli-Q water.

4.2. Tissue Collection

Tissue samples of a previous animal trial [13] (which had been stored frozen at −20 °C) were re-analyzed to focus on the accumulation, distribution, and persistence of foregut deamino metabolites of indospicine during and after cessation of the feeding period.

Briefly, in this study six camels (2–4 years) with a weight between 270–387 kg were fed with dried and chaffed *Indigofera spicata* to deliver 337 µg indospicine/kg BW/day for 32 days. The camels were sourced from the feral population in central Australia and purchased from a commercial supplier. Animal protocols for this study were approved by the Animal Ethics Committee of University of Queensland, Queensland, Australia (AEC Approval Number: SAFS/047/14/SLAI). The *Indigofera* plant material was fed daily in two equal meals at 9:00 am and 2:00 pm. On day 33, three of the animals (camels 1–3) were euthanized and tissues from 6 organs (muscle, heart, spleen, pancreas, liver, and kidney) were collected during necropsy and frozen (−20 °C) until needed.

To exclude previous exposure to indospicine (or metabolites), the plasma collection started 10 days before the experimental feeding phase. Venous blood samples were collected from the jugular vein from all six animals (camels 1–6) in weekly intervals during the treatment phase. This was continued with the remaining three animals (camels 4–6) in weekly intervals until day 76, and then in fortnightly intervals until the end of the experiment (day 132), 100 days after cessation of the *Indigofera* feeding. Blood was collected in lithium heparin containers and spun for 10 min at 4400 rpm at 19 °C (Sigma 4K10, Osterode am Harz, Germany), after which the plasma was collected and frozen at −30 °C until analysis.

4.3. Indospicine Extraction

The tissue samples were extracted as previously described [26]. Camel organ tissues were thawed; 0.5 g was mixed with 25 mL 0.1% HFBA and homogenized for 15 s using a Polytron T25 homogenizer (Labtek, Brendale, Australia). The homogenates were cooled for 20 min at 4 °C before centrifugation (4500 rpm, 10 min, 18 °C). One mL of the supernatant was spiked with the internal standard, and an aliquot of 450 µL was transferred into a pre-rinsed Amicon Ultra, 0.5 mL, 3 K centrifugal filter (Merck, Millipore, Kilsyth, VIC, Australia), which was centrifuged at 10,000 rpm for 20 min. The filtrate was transferred into an autosampler vial for UHPLC-MS/MS analysis.

Plasma samples were thawed and diluted 25 times with 0.1% HFBA. An aliquot of the diluted plasma sample was spiked with the internal standard, and 450 µL were transferred to a pre-rinsed Amicon Ultra centrifugal filter. The filtrate was transferred into an autosampler vial for UHPLC-MS/MS analysis.

4.4. UHPLC-MS/MS Analysis

Quantification of the compounds was done according to a previously validated UHPLC-MS/MS method, with modifications [26]. Separation and quantification were carried out on a Shimadzu Nexera X2 UHPLC system (Shimadzu, Rydalmere, NSW, Australia), combined with a Shimadzu LCMS-8050 triple quadrupole mass spectrometer, equipped with an electrospray ionization (ESI) source operated in positive mode. Indospicine, 2-aminopimelamic acid, and 2-aminopimelic acid were separated at

35 °C on an Acquity BEH C18 column (Waters, Rydalmere, NSW, Australia) (100 mm × 2.1 mm id, 1.7 μm) with 0.1% HFBA and 100% acetonitrile as mobile phases A and B, respectively. The flow rate was set to 0.3 mL/min and the injection volume was 2 μL. The following gradient was applied: 1–70% B (3 min), 70% B (isocratic for 1 min), 70–100% B (0.5 min), the total run time was 6 min. The interface temperature was set to 300 °C, and the heating block to 400 °C. Nitrogen was used as nebulizing gas (2.0 L/min) and drying gas (5.0 L/min), compressed air was used as heating gas (10.0 L/min), and argon was used as the CID gas maintained at 270 kPa.

Indospicine was quantified using stable isotope dilution assay with two MS/MS transitions for each compound, *m/z* 174.2 → 84.0 and *m/z* 174.2 → 111.1 for indospicine, and *m/z* 177.2 → 113.0 and *m/z* 177.2 → 114.1 for the internal standard D_3-L-indospicine.

External calibration curves were used for the quantification of 2-aminopimelamic acid and 2-aminopimelic acid. Specific SRMs were used for the identification: transition of *m/z* 175.00 → 112.05 (verified by transition of *m/z* 175.00 → 67.05) for 2-aminopimelamic acid, and transition of *m/z* 175.90 → 112.20 (verified by transition of *m/z* 175.90 → 69.15) for 2-aminopimelic acid. Unit resolution was used for both precursor and product *m/z* values. The collision energies (shown in Table 1) were optimized for each transition for maximum sensitivity.

Table 1. Collision energy for quantifier and verifying single reaction monitoring (SRM) transitions for compounds **1–4**.

Compound	Collison Energy (eV)	
	Quantifier SRM	Verifier SRM
Indospicine (**1**)	−23.0	−17.0
D_3-L-indospicine (**4**)	−17.0	−16.5
2-aminopimelamic acid (**2**)	−15.0	−28.0
2-aminopimelic acid (**3**)	−15.0	−20.0

Author Contributions: Study design, R.A.J., E.T.T.T. and M.T.F.; investigation and formal analysis, G.N., M.Y., C.G. and K.W.Y.; writing—original draft preparation, G.N.; writing—review and editing, M.T.F., R.A.J., C.G., and E.T.T.T.; project administration, M.T.F.; funding acquisition, M.T.F.

Funding: This study was funded by the University of Queensland and partly supported by the Academic Training Scheme for Institutions of Higher Education (SLAI) Scholarship (provided by Malaysian Government and Universiti Teknologi, Majlis Amanah Rakyat (MARA)) and Top-up Assistance Program Scholarship (TUAP) (provided by The University of Queensland).

Conflicts of Interest: The authors declare no conflict of interest.

References

1. Tan, E.T.T.; Materne, C.M.; Silcock, R.G.; D'Arcy, B.R.; Al Jassim, R.; Fletcher, M.T. Seasonal and species variation of the hepatotoxin indospicine in Australian *Indigofera* legumes as measured by UPLC-MS/MS. *J. Agric. Food. Chem.* **2016**, *64*, 6613–6621. [CrossRef]
2. Wilson, P.G.; Rowe, R. A revision of the Indigofereae (Fabaceae) in Australia. 1. *Indigastrum* and the simple or unifoliolate species of *Indigofera*. *Telopea* **2004**, *10*, 651–682.
3. Wilson, P.G.; Rowe, R. A revision of the *Indigofereae* (Fabaceae) in Australia. 2. *Indigofera* species with trifoliolate and alternately pinnate leaves. *Telopea* **2008**, *12*, 293–307. [CrossRef]
4. Wilson, P.G.; Rowe, R. Three new species of *Indigofera* (Fabaceae: Faboideae) from Cape York Peninsula. *Telopea* **2008**, *12*, 285–292. [CrossRef]
5. FitzGerald, L.M.; Fletcher, M.T.; Paul, A.E.H.; Mansfield, C.S.; O'Hara, A.J. Hepatotoxicosis in dogs consuming a diet of camel meat contaminated with indospicine. *Aust. Vet. J.* **2011**, *89*, 95–100. [CrossRef]
6. Fletcher, M.T.; Al Jassim, R.A.M.; Cawtdell-Smith, A.J. The occurrence and toxicity of indospicine to grazing animals. *Agriculture* **2015**, *5*, 427–440. [CrossRef]
7. Hegarty, M.P.; Kelly, W.R.; McEwan, D.; Williams, O.J.; Cameron, R. Hepatotoxicity to dogs of horse meat contaminated with indospicine. *Aust. Vet. J.* **1988**, *65*, 337–340. [CrossRef] [PubMed]

8. Hutton, E.M.; Windrum, G.M.; Kratzing, C.C. Studies on the toxicity of *Indigofera endecaphylla*: I. Toxicity for rabbits. *J. Nutr.* **1958**, *64*, 321–337. [CrossRef] [PubMed]
9. Nordfeldt, S.; Henke, L.A.; Morita, K.; Matsumoto, H.; Takahash, M.; Younge, O.R.; Willers, E.H.; Cross, R.F. Feeding tests with *Indigofera endecaphylla* Jacq. (Creeping indigo) and some observations on its poisonous effects on domestic animals. *Hawaii Agric. Exp. Station Coll. Agric. Univ. Hawaii Tech. Bull.* **1952**, *15*, 5–23.
10. Hegarty, M.P. Toxic amino acids in foods of animals and man. *Proc. Nutr. Soc. Australia* **1986**, *11*, 73–81.
11. Fowler, M.E. Plant poisoning in free-living wild animals: A review. *J. Wildl. Dis.* **1983**, *19*, 34–43. [CrossRef]
12. Tan, E.T.T.; Al Jassim, R.; D'Arcy, B.R.; Fletcher, M.T. Level of natural hepatotoxin (Indospicine) contamination in Australian camel meat. *Food Addit. Contam. Part A* **2016**, *33*, 1587–1595. [CrossRef] [PubMed]
13. Tan, E.T.T.; Al Jassim, R.; Cawdell-Smith, A.J.; Ossedryver, S.M.; D'Arcy, B.R.; Fletcher, M.T. Accumulation, persistence, and effects of indospicine residues in camels fed *Indigofera* plant. *J. Agric. Food. Chem.* **2016**, *64*, 6622–6629. [CrossRef] [PubMed]
14. Tan, E.T.T.; Al Jassim, R.; D'Arcy, B.R.; Fletcher, M.T. In vitro biodegradation of hepatotoxic indospicine in *Indigofera spicata* and its degradation derivatives by camel foregut and cattle rumen fluids. *J. Agric. Food. Chem.* **2017**, *65*, 7528–7534. [CrossRef] [PubMed]
15. Sultan, S.; Giles, C.; Netzel, G.; Osborne, S.A.; Netzel, M.E.; Fletcher, M.T. Release of indospicine from contaminated camel meat following cooking and simulated gastrointestinal digestion: Implications for human consumption. *Toxins* **2018**, *10*, 356. [CrossRef] [PubMed]
16. Tan, E.T.T.; Yong, K.W.L.; Wong, S.H.; D'Arcy, B.R.; Al Jassim, R.; De Voss, J.J.; Fletcher, M.T. Thermo-alkaline treatment as a practical degradation strategy to reduce indospicine contamination in camel meat. *J. Agric. Food. Chem.* **2016**, *64*, 8447–8453. [CrossRef]
17. Hegarty, M.P.; Pound, A.W. Indospicine, a hepatotoxic amino acid from *Indigofera spicata*: Isolation, structure, and biological studies. *Aust. J. Biol. Sci* **1970**, *23*, 831–842. [CrossRef]
18. Fletcher, M.T.; Reichmann, K.G.; Ossedryver, S.M.; McKenzie, R.A.; Carter, P.D.; Blaney, B.J. Accumulation and depletion of indospicine in calves (*Bos taurus*) fed creeping indigo (*Indigofera spicata*). *Anim. Prod. Sci.* **2018**, *58*, 568–576. [CrossRef]
19. Madsen, N.P.; Hegarty, M.P. Inhibition of rat liver homogenate arginase activity *in vitro* by the hepatotoxic amino acid indospicine. *Biochem. Pharmacol.* **1970**, *19*, 2391–2393. [CrossRef]
20. Biczó, G.; Hegyi, P.; Berczi, S.; Dósa, S.; Hracskó, Z.; Varga, I.S.; Iványi, B.; Venglovecz, V.; Wittmann, T.; Takács, T.; et al. Inhibition of arginase activity ameliorates L-arginine-induced acute pancreatitis in rats. *Pancreas* **2010**, *39*, 868–874.
21. Emmanuel, B. Urea cycle enzymes in tissues (liver, rumen epithelium, heart, kidney, lung and spleen) of sheep (*Ovis aries*). *Comp. Biochem. Physiol. B Biochem. Mol. Biol.* **1980**, *65*, 693–697. [CrossRef]
22. Morris, S.M. Regulation of enzymes of urea and arginine synthesis. *Annu. Rev. Nutr.* **1992**, *12*, 81–101. [CrossRef]
23. Hegarty, M.P.; Court, R.D. Indigofera spicata. In *Tropical Crops and Pastures Division of CSIRO 1975-1976 Annual Report*; Commonwealth Scientific and Industrial Research Organisation: Brisbane, QLD, Australia, 1976; p. 70.
24. Sultan, S.; Osborne, S.A.; Addepalli, R.; Netzel, G.; Netzel, M.E.; Fletcher, M.T. Indospicine cytotoxicity and transport in human cell lines. *Food Chem.* **2018**, *267*, 119–123. [CrossRef] [PubMed]
25. Lang, C.S.; Wong, S.H.; Chow, S.; Challinor, V.L.; Yong, K.W.L.; Fletcher, M.; Arthur, D.M.; Ng, J.C.; De Voss, J.J. Synthesis of L-indospicine, [5,5,6-$^2H_3$]-L-indospicine and L-norindospicine. *Org. Biomol. Chem.* **2016**, *14*, 6826–6832. [CrossRef]
26. Tan, E.T.; Fletcher, M.T.; Yong, K.W.; D'Arcy, B.R.; Al Jassim, R. Determination of hepatotoxic indospicine in Australian camel meat by ultra-performance liquid chromatography-tandem mass spectrometry. *J. Agric. Food. Chem.* **2014**, *62*, 1974–1979. [CrossRef] [PubMed]

toxins

MDPI

Article

Release of Indospicine from Contaminated Camel Meat following Cooking and Simulated Gastrointestinal Digestion: Implications for Human Consumption

Saira Sultan [1], Cindy Giles [2], Gabriele Netzel [1], Simone A. Osborne [3], Michael E. Netzel [1] and Mary T. Fletcher [1,*]

1 Queensland Alliance for Agriculture and Food Innovation (QAAFI), The University of Queensland, Health and Food Sciences Precinct, Coopers Plains, QLD 4108, Australia; saira.sultan@uq.net.au (S.S.); g.netzel@uq.edu.au (G.N.); m.netzel@uq.edu.au (M.E.N.)

2 Department of Agriculture and Fisheries, Queensland Government, Health and Food Sciences Precinct, Coopers Plains, QLD 4108, Australia; cindy.giles@daf.qld.gov.au

3 Commonwealth Scientific and Industrial Research Organisation, Agriculture and Food, St. Lucia, QLD 4067, Australia; simone.osborne@csiro.au

* Correspondence: mary.fletcher@uq.edu.au; Tel.: +61-7-3443-2479

Received: 25 May 2018; Accepted: 30 August 2018; Published: 3 September 2018

Abstract: Indospicine, a hepatotoxic arginine analog, occurs in leguminous plants of the *Indigofera* genus and accumulates in the tissues of grazing animals that consume these plants. Furthermore, indospicine has caused toxicity in dogs following consumption of indospicine-contaminated meat; however, the potential impact on human health is unknown. The present study was designed to determine the effect of simulated human gastrointestinal digestion on the release and degradation of indospicine from contaminated camel meat following microwave cooking. Results showed no significant ($p > 0.05$) indospicine degradation during cooking or in vitro digestion. However, approximately 70% indospicine was released from the meat matrix into the liquid digesta during the gastric phase (in the presence of pepsin) and increased to >90% in the intestinal phase (with pancreatic enzymes). Following human consumption of contaminated meat, this soluble and more bioaccessible fraction of intact indospicine could be readily available for absorption by the small intestine, potentially circulating indospicine throughout the human body to tissues where it could accumulate and cause detrimental toxic effects.

Keywords: indospicine; hepatotoxicity; meat; in vitro digestion; human

Key Contribution: Indospicine is released from naturally contaminated camel meat following cooking and in vitro human digestion but does not undergo significant breakdown. Therefore, hepatotoxic indospicine could be readily available for absorption across the small intestine following human consumption and digestion of naturally contaminated meat.

---

1. Introduction

The non-proteinogenic amino acid indospicine (L-6-amidino-2-amino-hexanoic acid) is a hepatotoxic arginine analog (Figure 1) found widely in plants of the *Indigofera* genus [1]. This genus contains over 700 species distributed across tropical Africa, Asia, Australia, and North and South America, and includes species such as *I. spicata*, *I. hendecaphylla*, *I. linnaei*, *I. lespedezioides*, *I. vicioides* and *I. volkensii* that have all been reported to contain in excess of 500 mg indospicine/kg dry matter of foliage. Indospicine has been found to be directly toxic to livestock [2], and to also act as a secondary toxin due to its unusual ability to accumulate in tissues of livestock grazing on these plants [3–5].

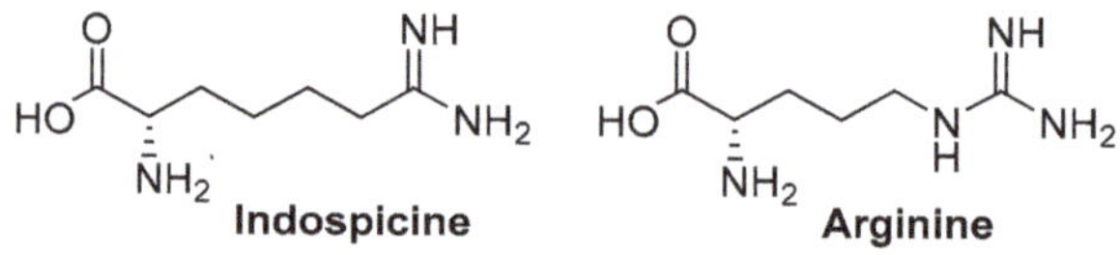

Figure 1. Chemical structures of the amino acids indospicine and arginine.

Indospicine hepatotoxicity has been demonstrated in rats following a single dose of indospicine that inhibited protein synthesis and induced fatty changes in, and enlargement of, the liver [6]. Livestock consuming *Indigofera* develop similar indospicine-induced liver hepatotoxicity; however, symptoms of toxicity frequently become apparent only after extended periods of *Indigofera* consumption [7]. Before the toxicity of *Indigofera* to livestock was recognized in the 1950s, several *Indigofera* species (notably *I. spicata* and *I. hendacphylla*) were introduced as pasture legumes in the USA and Australia. As a result, these species are now widely found in tropical regions far beyond their native range, across Africa, Asia, Australia, the Americas, and islands of both the Pacific and Indian Oceans [2]. *I. linnaei* and *I. lespedezioides* are also regionally abundant with extensive native ranges in Australia and South America respectively [2]. *Indigofera* plants are palatable legumes that are readily consumed by livestock leading to the reported indospicine accumulation in the meat of cattle [3], camels [4], and horses [5,8].

Indospicine is non-proteinogenic and instead occurs in both plant and animal tissues as a free amino acid [3–5]. It is a competitive inhibitor of arginase [9] and DNA synthesis [10], and has been shown to cause liver degeneration [11] and abortion [12]. In fact, indospicine hepatotoxicity has been demonstrated in all animal species investigated to date with reports of acute and sub-chronic hepatotoxic evidence for rats, [13] mice, [14] rabbits, [15] guinea pigs, [7] sheep, [7] dogs [5], and cattle [7]. However, the severity of the toxicity appears to vary considerably between species with dogs being particularly vulnerable [16]. Indeed, secondary poisoning of dogs has been reported in dogs consuming meat naturally contaminated with indospicine arising from livestock that had grazed on *Indigofera* plants [5,17].

With respect to digestion of indospicine following consumption of *Indigofera* plants, previous in vitro rumen studies have shown that indospicine can be metabolized by the rumenal microbial system [18]. However, little is known about the stability, release, and potential degradation of indospicine within monogastric digestive systems, such as the human gastro-intestinal tract, in which a single-chambered stomach secretes enzymes and acid to facilitate digestion prior to passage into the small and large intestines. When considering the digestive stability of one component, it is important to consider not only the chemical structure of a compound but also the nature of its bond to the food matrix. Therefore, static and/or dynamic in vitro digestion models that mimic the human gastrointestinal digestion process are a common approach to determine the matrix release (bioaccessibility) and stability of food components (like nutrients and toxins) as an initial measure to predict their potential bioavailability [19–22].

In the present study, an in vitro model was utilized to investigate stability, release, and degradation of indospicine to better understand the potential human exposure following the consumption of indospicine-contaminated meat. Specifically, the aims of this research were to investigate the effect of cooking (microwave) on indospicine stability in camel meat naturally containing indospicine, and most importantly, predict the bioaccessibility and digestive fate of indospicine using a static in vitro digestion model mimicking the human gastric and small intestinal digestion process. To date and to the authors' knowledge, this is the first study investigating the bioaccessibility of indospicine within a monogastric model system.

2. Results and Discussion

The hepatotoxic amino acid indospicine is found only in plant species of the *Indigofera* genus [1] and has been linked with poisoning of grazing livestock [2], and the deaths of dogs consuming contaminated meat from livestock that had grazed on these plants [5,17]. These canine deaths in Australia following the consumption of indospicine-contaminated camel [17] and horse meat [5] have raised both industry and consumer concern with regard to the potential contamination of meat for human consumption. The possible impact on human health is particularly concerning as all livestock grazing pastures that contain *Indigofera* plant species have the potential to accumulate indospicine as a free amino acid in their meat [3,4,23,24]. These indospicine residues are not readily excreted and can persist in tissues for up to 6 months after the cessation of *Indigofera* consumption, suggesting a strong affinity of indospicine with the meat matrix. Surveys of camel meat collected in Australia from animals killed in the field (in situ) and in abattoirs have demonstrated significant levels of indospicine residues above detectable limits (0.05 mg/kg), with meat from individual camels having levels up to 3.73 mg/kg [25]. Canine poisonings have occurred in the past following repeated consumption of indospicine-contaminated meat from the same or similar source repeatedly over several months [5,17]. The Australian supply of camel meat is sourced from more than 350,000 rangeland animals grazing arid inland regions of Australia [26], where *Indigofera* is also seasonally prevalent. These animals are slaughtered in abattoirs located within these regions before being supplied to growing niche markets in urban Australian Middle East and North African communities, as well as Central Australian local populations [26]. It is plausible that this lack of supply chain diversity could contribute to the same repeated exposure in these human consumers, and the potential health risk needs to be considered.

There are currently no identified studies providing epidemiological evidence or observational data indicating indospicine-induced adverse effects in humans. In the absence of such data, exposure risk can only be derived from available toxicity studies in dogs (often considered a model for human studies). The most substantial data available relates to thirteen different indospicine doses derived from two dog feeding experiments conducted for between 4 and 70 days [5,16]. A "lowest observed adverse effect level" (LOAEL) of 0.13 mg indospicine/kg bw/day is suggested based on observational data from the 70-day sub-chronic animal feeding experiment [5]. In this study, only minor histological liver lesions were observed in four dogs fed diets containing between 0.13 and 0.25 mg indospicine/kg bw/day [5]. A guidance value for human consumption of 1.3 μg indospicine/kg bw/day is thus proposed by dividing the selected LOAEL for dogs by an uncertainty factor of 10 to take into account mild degenerative changes to the liver in low dose dogs in the 70-day feeding trial, and by an additional factor of 10 to take into account intra species variation [27]. However, it can be postulated that a person with an average bodyweight of 70 kg [27] consuming the average Australian total daily meat intake of 143 g [28] could potentially consume 7.6 μg indospicine/kg bw/day if the dietary meat source was camel meat containing the reported 3.73 mg indospicine/kg [25]. This calculated intake of indospicine exceeds the proposed guidance value by a factor of 5. An additional factor of 10 could also be considered in the derived guidance value [27] due to the short duration of the literature study (70 days), and if this was implemented, then the calculated intake of indospicine could exceed the guidance value by a factor much higher than 5. Also, based on in vitro assessments of indospicine absorption using human intestinal cells [29], indospicine exhibits a 2-fold higher apparent permeability across an in vitro intestinal barrier compared to arginine (the amino acid analogue of indospicine). These findings indicate that indospicine is more readily absorbed than dietary arginine, suggesting preferential uptake that could potentiate further risks of toxicity.

Indospicine has been investigated in plant material by amino acid analyzer [5,30], high performance liquid chromatography (HPLC) with derivatization and UV detection [8,24] and liquid chromatography–tandem mass spectrometry (LC–MS/MS) [1,31]. However, the analysis of indospicine in meat is challenging due to low levels of contamination together with the complexity of the meat matrix. The incorporation of D_3-L-indospicine as an internal standard in sample extracts as used in this study can be beneficial in LC−MS/MS analysis as it overcomes the matrix effects observed in previous studies [17].

In the present study, indospicine-contaminated camel meat (microwave cooked) was subjected to in vitro digestion through sequential addition of pepsin in 0.1 M HCl and pancreatin-bile solutions with an appropriate adjustment of pH to mimic human in vitro digestion (Figure 2). Liquid and solid digesta from the in vitro gastric and small intestinal digestion were separated by centrifugation, prior to determination of indospicine concentration by LC-MS/MS (Figure 3), utilizing the previously reported and validated method [32]. Indospicine concentration was also measured in uncooked and cooked camel meat using the same method. All studies were carried out in triplicate with results shown in Figure 4 and expressed as the mean and standard deviation (SD). To enable a comparison of the liquid and solid phases, results are presented as indospicine content (μg) in each phase rather than concentration.

Results from this study indicate that there were no significant changes ($p > 0.05$) in the total indospicine content, suggesting that indospicine was not degraded during microwave cooking or gastrointestinal digestion in vitro (Figure 4). Cooking causes shrinkage of collagen fibres [33] and also increases meat protein surface hydrophobicity [34]. Indospicine is a water-soluble free amino acid, and this meat matrix breakdown during cooking and subsequent in vitro digestion resulted in an almost complete release of this amino acid from the solid phase into the liquid phase (Figure 4). This is evident from the observed release of approximately 70% indospicine from solid to liquid phase after the incubation of cooked meat with pepsin (during the gastric phase). Moreover, digestion with pancreatin and bile in the small intestinal phase resulted in a total release of more than 90% indospicine into the liquid digesta.

In contrast to the observed lack of indospicine degradation in the present in vitro model of gastrointestinal digestion in a monogastric system, indospicine was almost 100% degraded when *Indigofera* plant material was incubated in camel foregut fluid for 48 h [18]. This differing result is indicative of the presence of microbes able to degrade indospicine in the camel gastric system. Indospicine was similarly degraded when incubated with bovine ruminal fluid [18], and further studies are underway to isolate the responsible microbes with the potential to be utilized as a preventive probiotic. However, the observed accumulation of indospicine in camel tissues suggest that even though indospicine can be degraded by foregut fermentation, complete degradation does not occur before passage of the digesta into the intestine and a significant portion of indospicine is then available for absorption [18].

It must be noted that in vitro digestion models have several limitations that should be considered when interpreting the results. For example, no current in vitro model is capable of replicating all aspects of in vivo digestion, absorption, distribution, biodegradation (including the metabolic activity of the gut microbiota), and elimination [35,36]. Nevertheless, our results indicate that indospicine is released from the meat matrix and appears resistant to human gastrointestinal conditions, potentially making it available for absorption in the small intestine from liquid digesta. Postprandial indospicine may circulate throughout the human body to tissues and organs, such as the liver, where it could accumulate over time and cause detrimental, toxic effects.

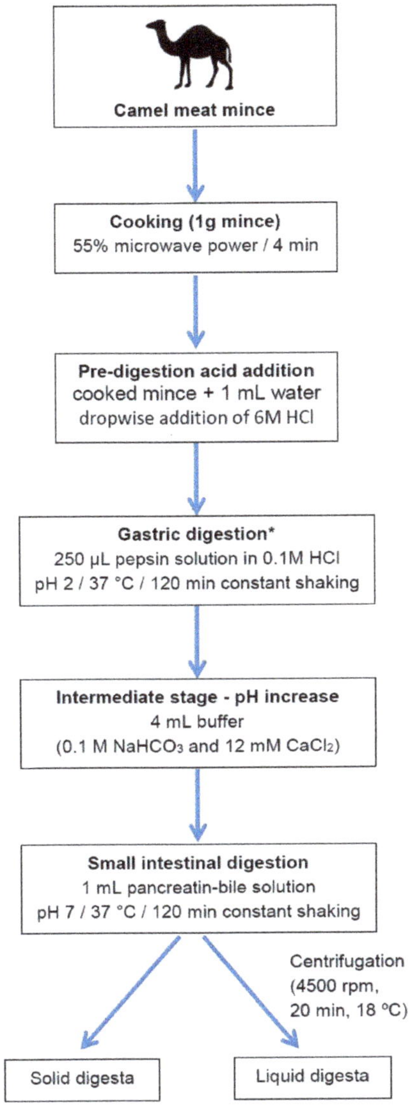

Figure 2. Schematic depiction of the stepwise in vitro digestion of camel mince under simulated human gastro-small intestinal conditions. (* Collection of gastric solid and liquid digesta as depicted for small intestinal digestion).

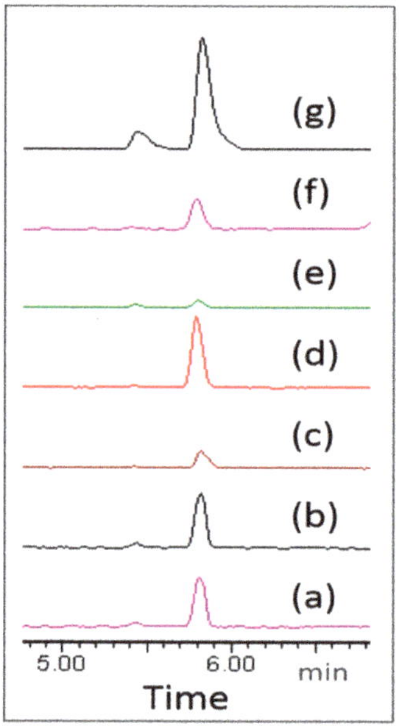

Figure 3. LC-MS/MS analysis of indospicine on Waters Micromass Quattro Premier triple quadrupole mass spectrometer utilizing a selected reaction monitoring (SRM) transitions of m/z 174 → 111 in (**a**) uncooked camel meat, (**b**) cooked camel meat, (**c**) solid phase of gastric digesta, (**d**) liquid phase of gastric digesta, (**e**) solid phase of small intestinal digesta, (**f**) liquid phase of small intestinal digesta, and (**g**) 0.05 mg/L standard indospicine solution.

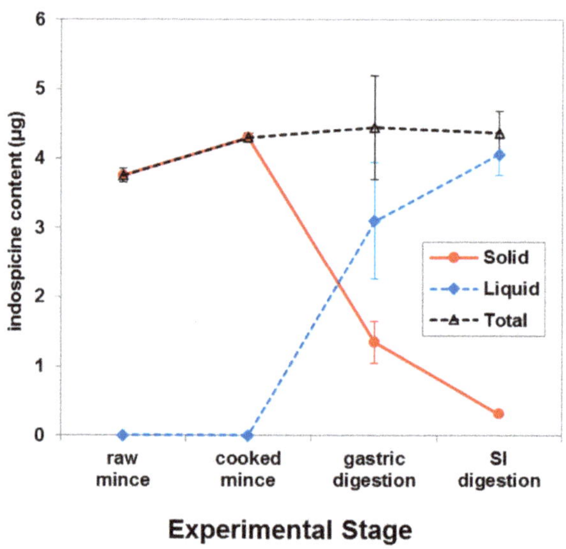

Figure 4. Indospicine content in raw, microwave cooked, and in vitro gastric and small intestinal (SI) digested indospicine-contaminated camel meat as determined by LC-MS/MS analysis. (All points are the mean of analysis of three replicates. Error bars show the SD).

There is no known mammalian enzyme that can degrade the amidino group of indospicine [37], and the preferred route to avoid indospicine toxicity is thus to prevent digestive uptake through degradation of the toxin during food processing. Our results suggest that the human digestive system does not have the capacity to degrade indospicine. Tan et al. [38] have recently reported that microwaving indospicine-contaminated camel meat under mild alkaline conditions (0.05% sodium bicarbonate, pH 8.8, 15 min) achieved 100% degradation of indospicine, with products identified as 2-aminopimelamic acid (major) and 2-aminopimelic acid (minor) (Figure 5). Such processing treatments may have ready applicability in the pet food industry given the recorded sensitivity of dogs to indospicine-contaminated meat, but are perhaps not appropriate in the processing of food for human consumption. Additionally, the metabolic fate and toxicity of indospicine hydrolysis products remains unknown and requires further investigation [38].

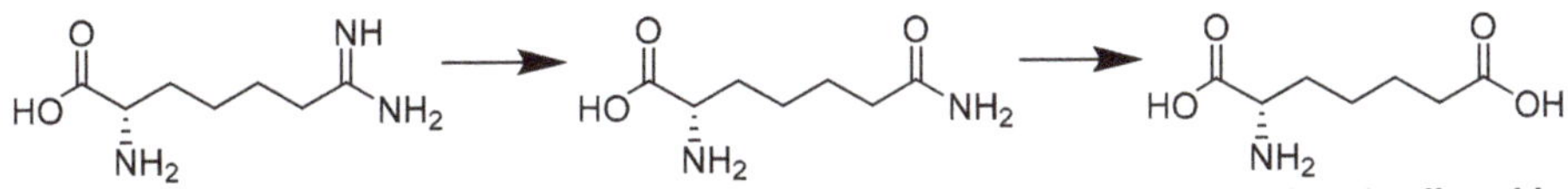

Figure 5. Hydrolysis of the amidino group of indospicine to corresponding amide (2-aminopimelamic acid) and acid (2-aminopimelic acid) under mild alkaline conditions as reported by Tan et al. [38].

3. Conclusions

Simple cooking of contaminated camel meat in a microwave, as carried out in the present study, does not degrade indospicine. Moreover, in vitro human gastrointestinal digestion conditions also had no effect on indospicine degradation and only helped to release the toxin from the solid meat matrix into the liquid digesta. These observations imply that following human consumption and digestion of contaminated meat, indospicine could be readily bioaccessible for absorption across the small intestine. The toxicity of indospicine for humans is uncertain [39], but the known toxicity in dogs (often considered a model for human studies) is particularly concerning. Camel meat is not commonly consumed by the broader Australian population, but is eaten by local indigenous populations and some immigrant ethnic groups within Australia. Further risk assessments, particularly for these high exposure groups, need to be undertaken with additional consideration given to possible indospicine contamination of other red meat supply chains.

4. Materials and Methods

4.1. Reagents

Unless otherwise stated, all chemicals were purchased from Sigma-Aldrich (Castle Hill, Sydney, NSW, Australia), and were of analytical or HPLC grade. De-ionized water was used throughout all experiments.

4.2. Study Design

In this study, cooked indospicine-contaminated camel meat (both meat and juices) was subjected to in vitro digestion through the sequential addition of pepsin and pancreatin-bile solutions with an appropriate adjustment of pH as outlined in Figure 2 to simulate the human gastro-small intestinal digestion process. All samples were digested in triplicate for both gastric digestion alone and for gastric plus small intestinal digestion. Liquid and solid digesta after gastric and small intestinal digestion were separated by centrifugation, and the concentration of indospicine in the digesta (liquid and solid) and uncooked and cooked camel meat was measured by LC-MS/MS.

4.3. Camel Meat Samples

Camel meat samples were obtained from a previously described experimental feeding trial in which camels were fed a diet containing the pasture legume *Indigofera spicata* with consequential accumulation of indospicine in meat tissues [3,4]. This feeding trial was conducted under approval of the Animal Ethics Committee of the University of Queensland, QLD, Australia (AEC Approval No. SAFS/047/14/SLAI; Date of approval: 19 March 2014). Indospicine-contaminated camel meat samples collected at autopsy [4] were utilized in this present study. Camel meat samples were minced using a commercial meat mincer (PRO 1400 meat grinder, Kenwood, Prestons, NSW, Australia) to provide a homogenous sample and stored frozen at −20 °C until used for further analysis.

4.4. In Vitro Digestion of Camel Meat

The in vitro digestion of camel meat samples was performed following the method described by Netzel et al. [36] with some modifications (Figure 2). The gastric phase was 120 min to account for reported variations in gastric emptying following consumption of meals that produce larger particle sizes like meat [40,41]. A 120 min intestinal phase was also employed.

Prior to in vitro digestion, camel meat (minced, 1 g each) in 15 mL screw–cap Falcon tubes was cooked in a microwave oven (Panasonic Genius NN5752–750 Watts, Sydney, NSW, Australia) for 4 min at medium heat (55% power, approximate temperature 70 °C).

4.4.1. Gastric Digestion

After cooking, samples were allowed to cool to room temperature and water (1 mL) was added to form a slurry. To prepare the samples for gastric digestion, the pH was lowered to 2.0 by the dropwise addition of HCl (6 M). To perform gastric digestion, 250 μL of pepsin solution (40 mg/mL pepsin from porcine gastric mucosa (1:2500 U/mg, Chem-Supply, Gillman, SA, Australia) dissolved in 0.1 M HCl) was added into the meat slurry and shaken manually to mix well. The mixture was incubated with continuous shaking at 37 °C for 120 min using an orbital mixer (RATEK Instruments, Boronia, VIC, Australia) placed in an incubator (Clayson IM550, Sydney, NSW, Australia). After 120 min constant shaking, tubes for gastric digestion were immediately centrifuged (4500 rpm, 20 min, 18 °C) to separate solid and liquid digesta.

4.4.2. Small Intestine Digestion

The sample tubes identified for small intestinal digestion were processed further. To these tubes, 4 mL of buffer containing 0.1 M $NaHCO_3$ and 12 mM $CaCl_2$ was added dropwise to slowly raise the pH to 5.7. The digesta samples were mixed well and incubated for a further 30 min at 37 °C under constant shaking. This intermediate step was integrated to mimic the transition from the gastric to the small intestine environment. To start the small intestinal digestion, the pH of the mixture was further raised to 7.0 by the dropwise addition of 1 M NaOH followed by the addition of 1 mL of pancreatin-bile solution (8 mg/mL pancreatin from porcine pancreas (102557, USP Grade, MP Biomedicals, LLC, Illkirch, France) and 12 mg/mL porcine bile extract (B8631, Sigma-Aldrich, St. Louis, MO, USA) in 0.1 M $NaHCO_3$). The digesta was again incubated at 37 °C for 120 min with constant shaking.

4.4.3. Separation of Liquid and Solid Digesta

After completion of each digestion step (gastric and small intestinal) sample tubes were centrifuged with a Sigma 6K15 centrifuge (Sigma Centrifuges, Osterode, Germany; 4500 rpm, 20 min, 18 °C) to separate solid and liquid digesta. Liquid and solid digesta were stored separately at −40 °C until extracted and analyzed for indospicine.

4.5. Preparation of External and Internal Standards for LC-MS/MS Analysis

Indospicine analysis of all samples was conducted by LC-MS/MS utilizing synthetic indospicine as an external standard for preparation of a calibration curve and deuterium-labelled D_3-L-indospicine as a stable isotopically labeled internal standard added to all samples and standard solutions to overcome matrix effects.

Synthesized indospicine as the external standard (>99% pure) and deuterium-labelled D_3-L-indospicine (>99% pure) as the internal standard were kindly provided by Dr. Robert Lang and Prof. James De Voss, School of Chemistry and Molecular Biosciences, The University of Queensland, St. Lucia, QLD, Australia [42]. Stock solutions for both internal and external standards were prepared in de-ionized H_2O with 0.1% heptafluorobutyric acid (HFBA) and were frozen at −20 °C until used. Internal (1 mg/L) and external (0.002–2 mg/L) standard solutions for indospicine LC-MS/MS quantification were prepared from the stock solutions and were stored frozen at −20 °C for no longer than a month before used.

4.6. Extraction of Indospicine from Camel Meat Samples and Digesta

Camel meat samples (uncooked, cooked, and both solid and liquid digesta) were extracted and analyzed by a previously validated and published liquid chromatography-tandem mass spectrometry method [32]. Prior to analysis, centrifugal filter units (Amicon® Ultra 0.5 mL 3K, Merck, Bayswater, VIC, Australia) were pre-rinsed and centrifuged (Microcentrifuge 5424, Eppendorf, North Ryde, NSW, Australia) at 10,000 rpm for 20 min with de-ionized water (2 × 300 µL) to remove glycerine, then inverted and spun for 1 min at 1000 rpm.

Minced un-cooked camel meat was thawed, weighed (0.5 g), and mixed with 0.1% HFBA (25 mL), followed by homogenization (Polytron T25, Labtek, Brendale, QLD, Brendale, Australia) for 15 s. The homogenized samples were chilled (4 °C) for 20 min and then centrifuged at 4500 rpm for 20 min at 18 °C. Aliquots of 1.0 mL of the resulting supernatants were spiked with 100 µL internal standard (D_3-L-indospicine, 1 mg/L in 0.1% HFBA), vortexed for 10 s, and a 450 µL portion was transferred into pre-rinsed centrifugal filters. The filtered sample mixture was then centrifuged (10,000 rpm, 20 min) and transferred to a limited volume insert (≈350 µL) for LC-MS/MS analysis.

Cooked meat and solid digesta were extracted and processed in a similar fashion to the raw meat. Liquid digesta (500 µL) was mixed with 0.1% HFBA (5 mL) and processed in a similar manner. All quantitations were calculated back to total indospicine content (µg) in either solid or liquid phase.

4.7. LC-MS/MS Analysis of Samples

Separation of the indospicine was achieved using a Waters ACQUITY UPLC® system (Waters, Lane Cove, NSW, Australia) equipped with a Waters BEH C18 column (1.7 µm, 100 mm length, 2.1 mm i.d.) at 30 °C and a flow rate of 0.2 mL/ min. The mobile phase was a mixture of (A) H_2O with 0.1% HFBA (*v*/*v*; pH 2.15) and (B) acetonitrile with 0.1% HFBA with the following gradient: 99% A to 70% A in 4 min, 70% A isocratic for 3 min, 70% A to 99% A in 1 min and 99% A for 2 min.

MS/MS detection was carried out using a Waters Micromass Quattro Premier triple quadrupole mass spectrometer with an electrospray ionisation (ESI) source operated in positive mode as previously described [32]. Eluted indospicine was quantitated utilizing selected reaction monitoring (SRM) transitions of m/z 174 → 111 (verified by transition of m/z 174 → 157) for indospicine, and m/z 177 → 114 (verified by transition of m/z 177 → 113) for D_3-L-indospicine as internal standard. The capillary voltage was 2.79 kV; cone gas flow was 50 L/h; desolvation gas flow was 600 L/h. The source and desolvation temperatures were set at 150 °C and 350 °C, respectively. Argon gas collision energy of indospicine (15 and 12 eV) and D_3-L-indospicine (15 and 15 eV) were set with cone voltage at 25 V.

4.8. Statistics

The data generated were processed using Microsoft Excel® 2010 (Microsoft, Redmond, WA, USA). Statistical analysis was conducted using ANOVA (GraphPad Prism, Version 6, La Jolla, CA, USA) with a completely randomized design. Differences were considered significant when p-values were below 0.05.

Author Contributions: Study design, M.T.F., S.A.O., G.N. and M.E.N.; Investigation and Formal Analysis, S.S. and C.G.; Writing-Original Draft Preparation, S.S.; Writing-Review & Editing, M.T.F., S.A.O., G.N., and M.E.N.; Project Administration, M.T.F.; Funding Acquisition, M.T.F.

Funding: This study was partly funded by an Endeavour International Postgraduate Scholarship sponsored by the Australian Government.

Conflicts of Interest: The authors declare no conflict of interest.

References

1. Tan, E.T.T.; Materne, C.M.; Silcock, R.G.; D'Arcy, B.R.; Al Jassim, R.; Fletcher, M.T. Seasonal and species variation of the hepatotoxin indospicine in Australian *Indigofera* legumes as measured by UPLC-MS/MS. *J. Agric. Food Chem.* **2016**, *64*, 6613–6621. [CrossRef] [PubMed]
2. Fletcher, M.T.; Al Jassim, R.A.M.; Cawdell-Smith, A.J. The occurrence and toxicity of indospicine to grazing animals. *Agriculture* **2015**, *5*, 427–440. [CrossRef]
3. Fletcher, M.T.; Reichmann, K.G.; Ossedryver, S.M.; McKenzie, R.A.; Carter, P.D.; Blaney, B.J. Accumulation and depletion of indospicine in calves (*Bos taurus*) fed creeping indigo (*Indigofera spicata*). *Anim. Prod. Sci.* **2018**, *58*, 568–576. [CrossRef]
4. Tan, E.T.T.; Al Jassim, R.; Cawdell-Smith, A.J.; Ossedryver, S.M.; D'Arcy, B.R.; Fletcher, M.T. Accumulation, persistence, and effects of indospicine residues in camels fed *Indigofera* plant. *J. Agric. Food Chem.* **2016**, *64*, 6622–6629. [CrossRef] [PubMed]
5. Hegarty, M.P.; Kelly, W.R.; McEwan, D.; Williams, O.J.; Cameron, R. Hepatotoxicity to dogs of horse meat contaminated with indospicine. *Aust. Vet. J.* **1988**, *65*, 337–340. [CrossRef] [PubMed]
6. Christie, G.S.; Madsen, N.P.; Hegarty, M.P. Acute biochemical changes in rat liver induced by the naturally-occurring amino acid indospicine. *Biochem. Pharmacol.* **1969**, *18*, 693–700. [CrossRef]
7. Nordfeldt, S.; Henke, L.A.; Morita, K.; Matsumoto, H.; Takahash, M.; Younge, O.R.; Willers, E.H.; Cross, R.F. Feeding tests with *Indigofera endecaphylla* Jacq. (Creeping indigo) and some observations on its poisonous effects on domestic animals. *Univ. Hawaii Agric. Exp. Stat. Technol. Bull.* **1952**, *15*, 5–23.
8. Ossedryver, S.M.; Baldwin, G.I.; Stone, B.M.; McKenzie, R.A.; Eps, A.W.; Murray, S.; Fletcher, M.T. *Indigofera spicata* (creeping indigo) poisoning of three ponies. *Aust. Vet. J.* **2013**, *91*, 143–149. [CrossRef] [PubMed]
9. Madsen, N.P.; Christie, G.S.; Hegarty, M.P. Effect of indospicine on incorporation of L-arginine-14C into protein and transfer ribonucleic acid by cell-free systems from rat liver. *Biochem. Pharmacol.* **1970**, *19*, 853–857. [CrossRef]
10. Christie, G.S.; De Munk, F.G.; Madsen, N.P.; Hegarty, M.P. Effects of an arginine antagonist on stimulated human lymphocytes in culture. *Pathology* **1971**, *3*, 139–144. [CrossRef] [PubMed]
11. Hegarty, M.P.; Pound, A.W. Indospicine, a hepatotoxic amino acid from *Indigofera spicata*: Isolation, structure, and biological studies. *Aust. J. Biol. Sci.* **1970**, *23*, 831–842. [CrossRef]
12. Pearn, J.H.; Hegarty, M.P. Indospicine—The teratogenic factor from *Indigofera spicata* extract causing cleft palate. *Br. J. Exp. Pathol.* **1970**, *51*, 34–36. [PubMed]
13. Christie, G.S.; Wilson, M.; Hegarty, M.P. Effects on the liver in the rat of ingestion of *Indigofera spicata*, a legume containing an inhibitor of arginine metabolism. *J. Pathol.* **1975**, *117*, 195–205. [CrossRef] [PubMed]
14. Hutton, E.M.; Windrum, G.M.; Kratzing, C.C. Studies on the toxicity of *Indigofera endecaphylla*: II. Toxicity for mice. *J. Nutr.* **1958**, *65*, 429–440. [CrossRef] [PubMed]
15. Hutton, E.M.; Windrum, G.M.; Kratzing, C.C. Studies on the toxicity of *Indigofera endecaphylla*: I. Toxicity for rabbits. *J. Nutr.* **1958**, *64*, 321–337. [CrossRef] [PubMed]

16. Kelly, W.R.; Young, M.P.; Hegarty, M.P.; Simpson, G.D. The hepatotoxicity of indospicine in dogs. In *Poisonous Plants*; James, L.F., Keeler, R.F., Bailey, E.M., Cheeke, P.R., Hegarty, M.P., Eds.; Iowa State University Press: Ames, IA, USA, 1992; pp. 126–130.
17. FitzGerald, L.M.; Fletcher, M.T.; Paul, A.E.; Mansfield, C.S.; O'Hara, A.J. Hepatotoxicosis in dogs consuming a diet of camel meat contaminated with indospicine. *Aust. Vet. J.* **2011**, *89*, 95–100. [CrossRef] [PubMed]
18. Tan, E.T.T.; Al Jassim, R.; D'Arcy, B.R.; Fletcher, M.T. In vitro biodegradation of hepatotoxic indospicine in *Indigofera spicata* and its degradation derivatives by camel foregut and cattle rumen fluids. *J. Agric. Food Chem.* **2017**, *65*, 7528–7534. [CrossRef] [PubMed]
19. Bobrich, A.; Fanning, K.J.; Rychlik, M.; Russell, D.; Topp, B.; Netzel, M. Phytochemicals in Japanese plums: Impact of maturity and bioaccessibility. *Food Res. Int.* **2014**, *65*, 20–26. [CrossRef]
20. Braga, A.C.; Alves, R.N.; Maulvault, A.L.; Barbosa, V.; Marques, A.; Costa, P.R. In vitro bioaccessibility of the marine biotoxin okadaic acid in shellfish. *Food Chem. Toxicol.* **2016**, *89*, 54–59. [CrossRef] [PubMed]
21. Ekbatan, S.S.; Sleno, L.; Sabally, K.; Khairallah, J.; Azadi, B.; Rodes, L.; Prakash, S.; Donnelly, D.J.; Kubow, S. Biotransformation of polyphenols in a dynamic multistage gastrointestinal model. *Food Chem.* **2016**, *204*, 453–462. [CrossRef] [PubMed]
22. Versantvoort, C.H.; Oomen, A.G.; Van de Kamp, E.; Rompelberg, C.J.; Sips, A.J. Applicability of an in vitro digestion model in assessing the bioaccessibility of mycotoxins from food. *Food Chem. Toxicol.* **2005**, *43*, 31–40. [CrossRef] [PubMed]
23. Hegarty, M.P. Non-metallic chemical residues in toxic plants with potential importance to animal and human health. In *Vet Update '92'*; Osborne, H.G., Ed.; University of Queensland. Continuing Professional Education: Brisbane, Australia, 1992; pp. 323–332.
24. Pollitt, S.; Hegarty, M.P.; Pass, M.A. Analysis of the amino acid indospicine in biological samples by high performance liquid chromatography. *Nat. Toxins* **1999**, *7*, 233–240. [CrossRef]
25. Tan, E.T.T.; Al Jassim, R.; D'Arcy, B.R.; Fletcher, M.T. Level of natural hepatotoxin (Indospicine) contamination in Australian camel meat. *Food Addit. Contam. Part A* **2016**, *33*, 1587–1595. [CrossRef] [PubMed]
26. Andrews, L.; Clarke, M.; Lethbridge, M.; Sobels, J. *Central Australian Commercial Camel Meat Viability Study, Report to the Northern Territory and South Australian Governments, 2015–2016. 71p*; Agriknowledge: Mylor, South Australia, 2016; Available online: https://agriknowledge2.weebly.com/agribusiness.html (accessed on 20 August 2018).
27. enHealth Council. Environmental Health Risk Assessment: Guidelines for Assessing Human Health Risks from Environmental Hazards. 2012. Available online: http://www.eh.org.au/documents/item/916 (accessed on 2 August 2018).
28. ABS. Australian Health Survey: Nutrition First Results—Foods and Nutrients, 2011-12. 2014. Available online: http://www.abs.gov.au/ausstats/abs@.nsf/detailspage/4364.0.55.0072011-12 (accessed on 2 August 2018).
29. Sultan, S.; Osborne, S.A.; Addepalli, R.; Netzel, G.; Netzel, M.E.; Fletcher, M.T. Indospicine cytotoxicity and transport in human cell lines. *Food Chem.* **2018**, *267*, 119–123. [CrossRef] [PubMed]
30. Aylward, J.; Haydock, K.; Strickland, R.; Hegarty, M. *Indigofera* species with agronomic potential in the tropics. Rat toxicity studies. *Crop Pasture Sci.* **1987**, *38*, 177–186. [CrossRef]
31. Gardner, D.R.; Riet-Correa, F. Analysis of the toxic amino acid indospicine by liquid chromatography-tandem mass spectrometry. *Int. J. Poisonous Plant Res.* **2011**, *1*, 20–27.
32. Tan, E.T.T.; Fletcher, M.T.; Yong, K.W.L.; D'Arcy, B.R.; Al Jassim, R. Determination of hepatotoxic indospicine in Australian camel meat by ultra-performance liquid chromatography–tandem mass spectrometry. *J. Agric. Food Chem.* **2014**, *62*, 1974–1979. [CrossRef] [PubMed]
33. Tornberg, E. Effects of heat on meat proteins—Implications on structure and quality of meat products. *Meat Sci.* **2005**, *70*, 493–508. [CrossRef] [PubMed]
34. Morita, J.-I.; Yasui, T. Involvement of hydrophobic residues in heat-induced gelation of myosin tail subfragments from rabbit skeletal muscle. *Agric. Biol. Chem.* **1991**, *55*, 597–599.
35. Hur, S.J.; Lim, B.O.; Decker, E.A.; McClements, D.J. In vitro human digestion models for food applications. *Food Chem.* **2011**, *125*, 1–12. [CrossRef]
36. Netzel, M.; Netzel, G.; Zabaras, D.; Lundin, L.; Day, L.; Addepalli, R.; Osborne, S.A.; Seymour, R. Release and absorption of carotenes from processed carrots (*Daucus carota*) using in vitro digestion coupled with a Caco-2 cell trans-well culture model. *Food Res. Int.* **2011**, *44*, 868–874. [CrossRef]

37. Hegarty, M.P. Toxic amino acids in foods of animals and man. *Proc. Nutr. Soc. Aust.* **1986**, *11*, 73–81.
38. Tan, E.T.T.; Yong, K.W.L.; Wong, S.-H.; D'Arcy, B.R.; Al Jassim, R.; De Voss, J.J.; Fletcher, M.T. Thermo-alkaline treatment as a practical degradation strategy to reduce indospicine contamination in camel meat. *J. Agric. Food Chem.* **2016**, *64*, 8447–8453. [CrossRef] [PubMed]
39. Pass, M.A. *Contaminated Horsemeat. Assessment and Prevention of Toxicity from Indospicine. RIRDC Report, Project No UQ-46A*; Rural Industries Research & Development Corporation: Kingston, ACT, Australia, 2000; p. 9.
40. Kong, F.; Singh, R.P. Disintegration of solid foods in human stomach. *J. Food Sci.* **2008**, *73*, R67–R80. [CrossRef] [PubMed]
41. Collins, P.J.; Horowitz, M.; Maddox, A.; Myers, J.C.; Chatterton, B.E. Effects of increasing solid component size of a mixed solid/liquid meal on solid and liquid gastric emptying. *Am. J. Physiol. Gastrointest. Liver Physiol.* **1996**, *271*, G549–G554. [CrossRef] [PubMed]
42. Lang, C.-S.; Wong, S.-H.; Chow, S.; Challinor, V.L.; Yong, K.W.L.; Fletcher, M.T.; Arthur, D.M.; Ng, J.C.; De Voss, J.J. Synthesis of L-indospicine, [5,5,6-$^{2}H_{3}$]-L-indospicine and L-norindospicine. *Org. Biomol. Chem.* **2016**, *14*, 6826–6832. [CrossRef] [PubMed]

toxins

MDPI

Article

Aflatoxin Contamination of Milk Marketed in Pakistan: A Longitudinal Study

Agha Waqar Yunus [1,*], Nida Imtiaz [1], Haider Khan [1], Mohammed Nawaz Mohammed Ibrahim [2] and Yusuf Zafar [3]

1 Animal Sciences Institute, PARC National Agricultural Research Center, Park Road, Islamabad 45500, Pakistan; nidaimtiaz234@yahoo.com (N.I.); haiderkhan_1@hotmail.com (H.K.)
2 International Livestock Research Institute, ILRI-Pakistan, Park Road, Islamabad 45500, Pakistan; m.ibrahim@cgiar.org
3 Pakistan Agricultural Research Council, G-5/1, Islamabad 44000, Pakistan; y_zafar@yahoo.com
* Correspondence: aghawaqaryunus@yahoo.com; Tel.: +92-51-90733931; Fax: +92-51-9255058

Received: 19 December 2018; Accepted: 1 February 2019; Published: 13 February 2019

Abstract: A longitudinal one-year study was conducted to determine aflatoxin M_1 levels in different types of milk marketed in Pakistan. Processed and raw liquid milk from 21 sources, two milk powder and six tea whitener brands were sampled on monthly basis from Islamabad. The aflatoxin M_1 levels in liquid milk were lower ($p < 0.05$) in summer (April to July) compared with the levels in winter (January, November and December). The mean aflatoxin M_1 levels were 254.9, 939.5, and 1535.0 ng/L in UHT, pasteurized, and raw milk, respectively (differing at $p < 0.001$). The mean toxin level in powdered milk after reconstitution was 522.1 ng/L. Overall, 12.9, 41.0, 91.9 and 50.0% of the UHT, pasteurized, raw and powdered milk samples, respectively, exceeded the Codex maximum tolerable limit of 500 ng of aflatoxin M_1/L. It was estimated that consumers of raw and processed milk were exposed to 11.9 and 4.5 ng aflatoxin M_1, respectively, per kg of body weight daily. The study indicates potential aflatoxin M_1 exposure risks for the consumers of raw milk in the country. The levels of the toxin though comparatively lower in milk powder, requires attention as this type of milk is consumed by infants.

Keywords: aflatoxin M_1; milk; pasteurized; Pakistan; UHT

Key Contribution: The mean AFM_1 levels in raw milk were found to be three times higher than the Codex maximum tolerable limit of 500 ng of AFM_1/L. The overall mean aflatoxin M_1 level in processed milk (UHT, pasteurized and powdered milk) was 63% lower than the raw milk, with only the UHT milk qualifying the Codex limit. Compared with the local milk powder brands, UHT milk could be used to reduce the daily aflatoxin M_1 exposure in infants.

1. Introduction

Aflatoxins are the toxic secondary metabolites of various *Aspergillus* spp. that commonly contaminate food and feed ingredients. The aflatoxins encountered in agricultural commodities include aflatoxin B_1, B_2, G_1 and G_2. In contaminated foodstuffs, the percentage of aflatoxin B_1 (AFB_1) in total aflatoxins is over 90%. Once ingested by animals, AFB_1 is also carried to milk in the form of the toxic metabolite aflatoxin M_1 (AFM_1) [1]. All of these toxins are known to exert potent hepatotoxic, immunotoxic and carcinogenic effects in animals and humans consuming the contaminated food. Due to high carcinogenicity, aflatoxins are the only group of mycotoxins for which legislation and control protocols are in place, even in many developing countries [2]. The toxicity of aflatoxins is known to be higher in younger age groups (infants, children and young animals). Monitoring the levels of AFM_1 in milk and baby foods is therefore more critical. Consequently, the levels of the mycotoxin allowed in milk are lower than the levels allowed in other foodstuffs. The EU further restricts the levels allowed in infant milk formula to half of the levels allowed in milk.

The maximum tolerable limit of AFM_1 in liquid milk is 500 ng/L in the USA and in the Codex standards, while only 50 ng/L in the EU. In Pakistan, the maximum tolerable limit of AFM_1 is 10 μg/kg in milk powder while no particular legislation has been made for liquid milk. This is despite the fact that specific monsoon conditions in the country favor mycotoxin development in food and feedstuffs, pushing Pakistan into a high risk area. The studies conducted in Pakistan also show that 25 to 90% of milk samples [3–6] could be contaminated with AFM_1.

There have been notable differences in the AFM_1 levels in milk reported by different authors from Pakistan. In this regard, Muhammad et al. [4] reported 17,380 ng/L as the mean AFM_1 level in milk sampled from Lahore in the year 2007, with 81% samples exceeding the 500 ng AFM_1/L limit. Contrary to this, Iqbal et al. [6] reported 64 ng/L mean AFM_1 level in milk sampled in the year 2011 in the urban areas of Punjab province, with 15% samples exceeding the 500 ng AFM_1/L limit. These differences in the AFM_1 contamination level reported by various authors could be due to different seasons, different feeds used by farmers in different areas, and different methods of AFM_1 quantification. Overall, such differences make it impractical to infer risk of exposure for the consumers of milk in other cities. The present study was therefore conducted as a longitudinal one year study to determine the AFM_1 levels in various types of milk, primarily processed, available in Islamabad the capital city of Pakistan. To the best of our knowledge, there is no previous longitudinal study on AFM_1 contamination in processed milk in Pakistan. Also, AFM_1 contamination of milk has not been previously investigated in Islamabad city. Data on processed milk from one city are however applicable to milk consumers in other cities because processing companies collect milk from farmers located in different areas and distribute it to consumers in all cities of Pakistan.

2. Results and Discussion

2.1. *Aflatoxin Contamination of Liquid Milk*

The seasonal trend regarding aflatoxin contamination of liquid milk (average of UHT, pasteurized, and raw milk) is presented in Figure 1. The levels of AFM_1 in liquid milk were lower ($p \leq 0.021$) in the summer months of April to June compared with the levels in the winter months of January, November and December. This seasonal variation in AFM_1 levels is in accordance with previous findings in which comparatively lower levels of AFM_1 were found in raw milk sampled during summer months in the country [7,8].

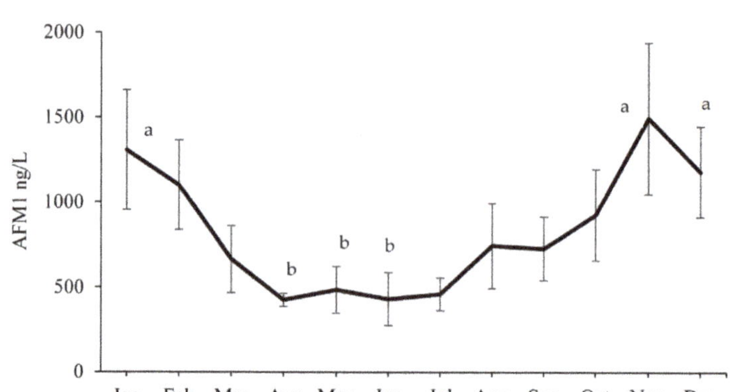

Figure 1. Average aflatoxin M_1 levels ± SEM, in liquid milk during different months (means bearing different alphabets differ significantly at $p < 0.05$).

The data regarding AFM_1 levels in different types of liquid milk are presented in Table 1. In case of UHT milk, the mean AFM_1 level was 254.9 ng/L, with 12.9% samples exceeding 500 ng/L, the maximum allowed limit in the USA. The maximum AFM_1 level noted was 1536 ng/L. Only five out of 85 UHT milk samples (5.8%) had AFM_1 levels less than 50 ng of AFM_1/L, the maximum allowed limit in the EU. Overall, the UHT milk samples had lower ($p < 0.001$) AFM_1 levels compared with the pasteurized and raw milk samples.

Table 1. Aflatoxin M_1 levels (ng/L) [1] in liquid milk samples during the year 2016.

Month	UHT Milk	Pasteurized Milk	Raw Milk	*p* Value	*F* Value
Jan	434.0 ± 517.5 [b]	1861.1 ± 1855.9 [ab]	2422.3 ± 198.6 [a]	0.043	4.12
Feb	135.5 ± 34.4 [b]	1613.9 ± 905.0 [a]	2155.7 ± 603.9 [a]	0.000	18.25
Mar	48.3 ± 37.8 [c]	556.0 ± 476.5 [b]	1908.7 ± 367.3 [a]	0.000	40.32
Apr	413.2 ± 201.2	334.0 ± 151.6	549.2 ± 38.3	0.055	3.49
May	109.5 ± 48.2 [b]	553.8 ± 770.5 [ab]	994.6 ± 539.7 [a]	0.029	4.41
Jun	61.4 ± 26.1 [b]	355.2 ± 351.1 [b]	1123.3 ± 1015.2 [a]	0.022	5.07
Jul	213.5 ± 150.4	565.3 ± 666.6	628.3 ± 134.4	0.169	1.98
Aug	218.9 ± 82.1	1148.3 ± 1820.0	886.2 ± 289.0	0.289	1.34
Sep	278.2 ± 124.1	813.8 ± 1073.7	1365.9 ± 459.2	0.078	3.04
Oct	279.5 ± 99.5 [b]	913.3 ± 1238.3 [b]	2231.7 ± 847.1 [a]	0.008	6.89
Nov	441.1 ± 184.8 [b]	1385.6 ± 1469.5 [b]	3796.3 ± 2858.1 [a]	0.009	6.40
Dec	395.1 ± 155.5 [b]	1570.9 ± 1457.5 [a]	1894.4 ± 1184.4 [a]	0.040	3.92
n	85	78	55		
Mean	254.9 ± 223.9 [c]	939.5 ± 1164.6 [b]	1535.0 ± 1234.5 [a]	0.000	31.74
Median	198.4	368.5	1037.6	-	-
Max	1536.0	4808.0	7460.7	-	-
Min	n.d.	32.8	1912	-	-

[1] Arithmetic means ± standard deviation. [abc] Means bearing different superscripts differ significantly within a row ($p < 0.05$). n.d. = less than detection limit.

In case of pasteurized milk, the mean AFM_1 level was 939.5 ng/L and only 41% samples were below the limit of 500 ng AFM_1/L. Only four pasteurized milk samples (5.1%) were below the limit of 50 ng AFM_1/L. Compared with the UHT milk, the pasteurized milk had higher ($p < 0.001$) AFM_1 levels. Raw milk was found to have higher ($p < 0.001$) AFM_1 levels than the UHT and pasteurized milk. None of the 34 raw milk samples qualified the EU limit of 50 ng AFM_1/mL milk. Only five raw milk samples (9.1%) had AFM_1 levels lower than the USA standard of 500 ng/L of milk, indicating potential health risks for the consumers of raw milk. Mean AFM_1 level in raw milk in the country has been reported to be in the range of 46 to 340 ng/L by various authors, which is lower than the mean level being reported here [5,8,9]. However, the mean AFM_1 level in raw milk collected from peri-urban farms in Lahore was found in one report to be 17,380 ng/L which is very high compared to the present findings [4]. The differences in various reports indicate that the AFM_1 contamination may vary in different seasons and areas.

In general, the milk sampled from rural areas of Pakistan, where more green fodder is available, has been found to have lower AFM_1 levels compared to the milk collected from urban areas [6]. This could also explain the lower levels found in UHT milk in the present study which is in general collected from rural areas followed by testing for AFM_1 levels before packaging. The farmers are usually paid according to the quality of the milk. The UHT milk companies also have the option to either not buy or discard the milk having higher AFM_1 contamination. Compared to the UHT milk, the companies marketing pasteurized milk have their own dairy farms and thus do not have the option to discard milk. It is therefore important to educate these companies to control aflatoxin in the dairy feeds, coupled with legislation, for a market advantage.

2.2. *Aflatoxin Contamination of Tea Whiteners and Milk Powder*

The data regarding AFM_1 levels in tea whiteners are presented in Table 2. The mean AFM_1 level in tea whiteners was only 98.9 ng/L, with 2.1% samples (1 out of 48) exceeding the 500 ng AFM_1/L limit. The mean AFM_1 levels in different months did not differ statistically (p = 0.081). Tea whiteners are usually made using both the milk and non-milk constituents, hence it is logical to expect a lower AFM_1 contamination in this commodity. It should however be noted that tea whiteners are not a substitute of milk, and the present results do not imply use of such products as a replacement of milk.

Table 2. Aflatoxin M_1 levels (ng/L) in tea whiteners.

Month	Tea Whiteners
Jan	138.8 ± 195.7
Feb	49.4 ± 83.3
Mar	0.6 ± 0.9
Apr	178.2 ± 173.5
May	16.4 ± 6.2
Jun	32.1 ± 24.4
Jul	69.4 ± 89.3
Aug	120.2 ± 55.4
Sep	60.9 ± 40.7
Oct	38.5 ± 30.5
Nov	120.8 ± 80.8
Dec	339.4 ± 357.3
p value	0.081
F value	1.849
n	48
Overall Mean	98.9 ± 161.8
Median	40.6
Max	932.6
Min	n.d.

Data presented as arithmetic means ± standard deviation; n.d. = less than detection limit.

The mean AFM_1 level in powdered milk after reconstitution was 522.1 ng/L i.e., close to the maximum tolerable limit of 500 ng AFM_1/L. In total, 50.0% of the powdered milk samples had AFM_1 levels lower than the 500 ng/L limit. Only two powder milk samples (10.0%) passed the EU standard of 50 ng AFM_1/L. It appears from these results that the powdered milk was a safer option for consumers compared to the pasteurized and raw milk. However, these results need to be examined in consideration of the fact that the studied milk powder brands are used as infant milk formula. As infants are more sensitive to the hazards of AFM_1, effective measures should be in place to ensure that all powdered milk brands marketed in Pakistan meet at least the 500 ng AFM_1/L limit.

2.3. *Miscellaneous Quality Variables of Milk Samples*

Miscellaneous quality variables were studied to identify any case of adulteration which would result in dilution of the natural AFM_1 content of milk and making comparison biased. Data regarding fat, total solids, and SNF content of milk samples are presented in Table 3. Fat percentage was lower ($p \leq 0.029$) in the UHT milk compared with the pasteurized milk samples in the month of January. The UHT milk samples were also found to have lower ($p \leq 0.037$) total solids compared with both the pasteurized and raw milk samples in January. These data are indicative of fat removal by milk processors to make milk uniform regarding fat content. The SNF contents of different types of milk did not differ during the study duration. Furthermore, no addition of starch, cane sugar, urea, detergent and hydrogen peroxide was found in liquid milk samples. These data indicate that no particular adulteration occurred in the liquid milk samples included in the study.

Table 3. Quality variables of liquid milk included in the study.

Mon	Fat (%)				Total Solids (%)				SNF (%)			
	UHT	Pasteur	Raw	p	UHT	Pasteur	Raw	p	UHT	Pasteur	Raw	p
Jan	3.21 ± 0.35^{b}	4.12 ± 0.13^{a}	3.84 ± 0.69^{a}	0.014	11.00 ± 1.22^{b}	12.56 ± 0.18^{a}	12.22 ± 1.02^{a}	0.027	7.78 ± 0.96	8.44 ± 0.23	8.38 ± 1.46	0.498
Feb	3.49 ± 0.29	4.12 ± 0.13	3.86 ± 0.64	0.077	11.62 ± 0.93	12.50 ± 0.28	11.79 ± 1.39	0.361	8.13 ± 0.69	8.37 ± 0.31	7.93 ± 1.48	0.758
Mar	3.66 ± 0.38	4.58 ± 0.13	3.90 ± 0.90	0.061	11.73 ± 0.87	12.31 ± 0.34	12.11 ± 0.62	0.331	8.07 ± 0.78	7.72 ± 0.37	8.22 ± 1.03	0.591
Apr	3.58 ± 0.23	3.72 ± 0.38	3.59 ± 0.69	0.868	11.66 ± 0.63	12.00 ± 0.45	11.83 ± 1.18	0.808	8.08 ± 0.57	8.27 ± 0.26	8.25 ± 1.24	0.911
May	3.36 ± 0.46	3.90 ± 0.28	4.09 ± 0.85	0.103	11.49 ± 0.98	12.00 ± 0.72	12.53 ± 0.89	0.106	8.12 ± 0.61	8.08 ± 0.51	8.44 ± 1.45	0.778
Jun	3.46 ± 0.27	3.76 ± 0.50	4.01 ± 0.94	0.302	11.73 ± 0.68	12.21 ± 0.62	12.26 ± 0.87	0.364	8.24 ± 0.44	8.45 ± 0.28	8.25 ± 1.29	0.914
Jul	3.63 ± 0.63	3.96 ± 0.44	4.14 ± 1.06	0.484	11.78 ± 1.16	12.37 ± 0.35	12.21 ± 1.04	0.547	8.15 ± 0.71	8.40 ± 0.33	8.07 ± 1.37	0.838
Aug	3.36 ± 0.18	3.10 ± 0.97	3.88 ± 0.61	0.100	11.58 ± 0.85	11.25 ± 1.33	11.97 ± 1.15	0.513	8.21 ± 0.67	8.15 ± 0.53	8.10 ± 1.41	0.978
Sep	3.56 ± 0.32	4.00 ± 0.45	4.01 ± 0.91	0.336	11.84 ± 0.93	12.53 ± 0.55	12.19 ± 1.34	0.548	8.38 ± 0.65	8.53 ± 0.23	8.15 ± 1.68	0.848
Oct	3.52 ± 0.22	3.96 ± 0.53	3.97 ± 1.14	0.494	12.59 ± 0.70	13.15 ± 0.78	12.56 ± 1.12	0.484	9.07 ± 0.60	9.19 ± 0.25	8.58 ± 1.28	0.438
Nov	3.54 ± 0.29	4.00 ± 0.71	4.16 ± 0.80	0.193	11.18 ± 1.55^{b}	12.98 ± 0.85^{a}	12.10 ± 0.72^{ab}	0.040	7.64 ± 1.41	9.00 ± 0.24	7.93 ± 0.87	0.090
Dec	3.41 ± 0.34	3.76 ± 0.32	4.04 ± 0.63	0.069	11.46 ± 1.18	11.81 ± 0.56	12.04 ± 0.79	0.474	8.04 ± 0.99	8.05 ± 0.49	8.01 ± 1.19	0.996

Data presented as arithmetic means ± standard deviation; UHT = milk with ultraheat treatment, Pasteur. = pasteurized milk. [abc] Means bearing different superscripts differ significantly within a row under each quality variable ($p < 0.05$).

In the case of tea whiteners, one brand was found to be positive for neutralizer/soda in April and then from June through December. Another brand of tea whitener was found to be positive for detergent only in January.

2.4. Exposure to Aflatoxin through Milk

The users of raw and processed milk (UHT and pasteurized) were estimated to be exposed to 11.9 and 4.5 ng AFM_1 per kg body weight daily, respectively (Table 4). The overall daily AFM_1 intake was found to be 11.6 ng per kg body weight, which is very high compared with 3.1 ng/kg mean estimated previously for consumers in Karachi [10]. The higher exposure estimated in the present study could be due to the higher contamination levels noted in Islamabad, and the use of weighted AFM_1 mean in our study. The mean AFM_1 in milk if not corrected for relative consumption would have resulted in an estimate of 6.4 ng/kg day.

Table 4. Estimated daily AFM_1 intake in users of different types of milk.

Type of Milk	AFM_1 (ng/L)	Daily Milk Consumption [1] (mL)	Average Body Weight [2] (kg)	Daily Intake (ng/kg bw)
Children:				
Milk powder for 1 year old	353.1	500	9.3	19.1
Milk powder for 5 year old	728.7	500	18.5	19.7
Adults:				
UHT	254.9	466	60.0	2.0
Pasteurized	939.5	466	60.0	7.3
Raw	1535.0	466	60.0	11.9
Overall liquid	1487.4	466	60.0	11.6

[1] Milk consumption for children as recommended by the milk powder manufacturer. Milk consumption in case of adult is the per head milk available in Pakistan [11]. [2] Average body weight of children taken from Royal College of Pediatrics and Child Health (UK-WHO) charts. Average weight of Pakistani children of one year and five years of age are 5.6 and 15.3, respectively [12]. bw = body weight.

The presently estimated AFM_1 intake for children using milk powder is alarmingly high despite the acceptable levels of AFM_1 in milk powder for one-year-old children. Compared to adults, the consumption of milk per kg body weight is higher in case of children which results in higher contribution of aflatoxin from milk. It may be noted that UHT milk was found to have the lowest AFM_1 levels in the present study. If UHT milk was used for calculation of daily exposure of children, then the figures for one- and five-year-old children had been 13.8 and 6.9 ng per kg body weight. This shows the need to reduce the levels of AFM_1 in milk powder produced in the country, especially in the milk for the five-year age group. In a separate study (unpublished data), we found that the AFM_1 levels in imported milk powders was less than 60 ng/L and would result in an exposure of only 3.2 ng/kg body weight daily in one-year-old children. This implies that during legislation, the levels of AFM_1 allowed in milk for one-year-old and younger children should be kept lower than 500 ng/L.

3. Conclusions

This study, though conducted on a limited scale, identifies raw milk as less suitable for human consumption in the capital city Islamabad due to high AFM_1 levels. The processed liquid milk was found to be a safer option for consumers. In all types of milk, the levels of AFM_1 were lower during summer months. Reducing the levels of AFM_1 in milk for infants, while strictly following the 500 ng AFM_1/L limit in milk for older children are suggested as ways to reduce the exposure of children to AFM_1.

4. Materials and Methods

4.1. Sampling

All the regularly available locally produced processed milk brands were collected once a month from supermarkets in the twin cities of Islamabad-Rawalpindi in the year 2016. Each time, 21 different liquid milk samples were collected including eight UHT milk brands, seven pasteurized milk brands, and raw milk from six different suppliers. Additionally, two brands of milk powder—each for a different age group of children—and six brands of tea whiteners were sampled on monthly basis.

As UHT and powdered milk have long expiry time, it was ensured that a different and fresh batch was collected on each sampling. The collected liquid milk samples were transported in chilled form to the laboratory where these were subsampled and stored at −20 °C until AFM_1 analysis. Identity of the collected samples was masked from analysts to make the study blind.

4.2. Aflatoxin M_1 Analysis

The milk samples were analyzed for AFM_1 contamination using ELISA kits (AM1-E01, Immunolab GmbH, Kassel, Germany) following protocols specified by the manufacturer. The kit had a quantification range of 10 to 1000 ng of AFM_1/L. Samples were analyzed several times in different dilutions until the AFM_1 levels in the diluted samples fell within the quantification range of the ELISA kit. Limit of detection (LOD) was found to be 4.4 ng AFM_1/L, while recovery of AFM_1 was 86.9% at 500 ng of AFM_1/L. Results were not corrected for recovery.

The kit was validated before start of this study [13]. In addition, external standards were run with each microtiter plate to ensure reliability of the AFM_1 quantification. Two types of reference materials were used to make external standards. The first was a 44 ng of AFM_1/L reference skim milk powder (RMBD-248, EU Joint Res Center, IRMM, Geel, Belgium) which was used to make 4.4 and 44 ng of AFM_1/L standards. The second reference material was a purified 9.786 µg of AFM_1/mL solution (46319U, Supelco, Bellefonte, PA, USA) which was diluted with ddH_2O to make 5, 50, 500 ng of AFM_1/L secondary standards for use with each microtiter plate.

The ELISA kits were read on an ELISA reader (BDSL Immunoskan MS 355, Labsystems, Vantaa, Finland). For quantification of AFM_1, a software based on four parametric curve estimation provided by the kit manufacturer was used.

4.3. Miscellaneous Milk Quality Variables

Raw milk in the country is often adulterated with water, urea, starch, and neutralizers, etc [14]. Comparison of adulterated milk with pure milk is not justified because addition of water and non-milk constituents result in milk dilution and lowering of the AFM_1 contents. Therefore, all the liquid milk samples were analyzed for milk fat, total solids, and solids-not-fat (SNF) to identify any notable case of adulteration. Milk fat was determined following the Gerber method and using a Gerber Centrifuge (Funke Dr. N. Gerber, Berlin, Germany) [15]. Total solids were determined using Quevenne lactometer and following the principles of lactodensitometry as detailed by Ling [16]. Solids-not-fat were calculated as difference between total solids and milk fat.

Qualitative tests were performed to detect adulterants and stabilizers including starch, cane-sugar, urea, detergent, and hydrogen peroxide following methods described by Sharma et al. [17].

4.4. Calculations and Statistics

The exposure to AFM_1 through a specific milk type was estimated using a formula:

$$\begin{gathered}\text{Average daily } AFM_1 \text{ intake (ng/kg body weight) through a specific type of milk} \\ = [(\text{mean } AFM_1 \text{ ng/L in the specific type of milk}) \\ \times (\text{milk availability per head per day in Pakistan})]/(\text{average body weight})\end{gathered} \tag{1}$$

where, milk availability was taken as 170.2 L/head/year [11], and average body weight was taken as 60 kg. Daily consumption of milk powder in case of children was taken as recommended by the milk powder manufacturer. To calculate the exposure level for one and five year old children, average body weight was taken from Royal College of Pediatrics and Child Health (UK-WHO) charts.

In case of overall AFM_1 intake through liquid milk, the mean AFM_1 levels were normalized using a formula, keeping in view only 5% share of processed milk in the overall milk market:

$$\begin{aligned}&\text{Average daily } AFM_1 \text{ intake through liquid milk (ng/kg body weight)}\\&= [(\text{mean } AFM_1 \text{ in raw milk} \times 95) + (\text{mean } AFM_1 \text{ in processed milk} \times 5)]/100\end{aligned} \quad (2)$$

Means were compared using ANOVA and LSD in IBM SPSS Statistics 20 (IBM Corp., Armonk, New York, NY, USA, 2011). Values of AFM_1 less than the LOD were considered as zero for calculating monthly means.

Author Contributions: Conceptualization, A.W.Y., and M.N.M.I.; Methodology, A.W.Y. and H.K.; Aflatoxin M_1 Analysis, N.I. and A.W.Y.; Milk Quality Analyses, H.K.; Data Curation, A.W.Y.; Writing-Original Draft Preparation, A.W.Y.; Writing-Review & Editing, A.W.Y., M.N.M.I. and Y.Z.; Visualization, A.W.Y., M.N.M.I. and Y.Z.; Supervision, A.W.Y. and M.N.M.I.; Project Administration and Funding Acquisition, A.W.Y. and M.N.M.I.

Funding: This research was funded under the framework of collaborative research agreement between National Agricultural Research Center (NARC) and International Livestock Research Institute (ILRI) under the USAID funded Agriculture Innovation Program (AIP) for Pakistan led by CIMMYT. The findings of this study are the sole responsibility of NARC and ILRI, and do not necessarily reflect the views of USAID or United States Government.

Acknowledgments: The authors highly acknowledge the support given by Aman Ullah, Umer Farooq, Eitazaz Ehsan, Zahid and other staff members of various laboratories of Animal Sciences Institute, PARC National Agricultural Research Center, Islamabad during analyses of milk samples. Also, the contribution of Johanna Lindhal (ILRI scientist) for her editorial comments is appreciated.

Conflicts of Interest: The authors declare no conflict of interest.

References

1. Yunus, A.W.; Razzazi-Fazeli, E.; Böhm, J. Aflatoxin B1 in affecting broiler's performance, immunity, and gastrointestinal tract: A review of history and contemporary issues. *Toxins* **2011**, *3*, 566–590. [CrossRef] [PubMed]
2. Food and Agriculture Organization (FAO). *Worldwide Regulations for Mycotoxins in Food and Feed in 2003*; FAO Food and Nutrition Paper 81; Food and Agriculture Organization of the United Nations: Rome, Italy, 2004.
3. Raza, R. Occurrence of aflatoxin M_1 in the milk marketed in the city of Karachi, Pakistan. *J. Chemical Soc. Pakistan* **2006**, *28*, 155–157.
4. Muhammad, K.; Tipu, M.Y.; Abbad, M.; Khan, A.M.; Anjum, A.A. Monitoring of aflatoxin M_1 in market raw milk in Lahore City, Pakistan. *Pakistan J. Zool.* **2010**, *42*, 697–700.
5. Iqbal, S.Z.; Asi, M.R.; Ariño, A. Aflatoxin M_1 contamination in cow and buffalo milk samples from the North West Frontier Province (NWFP) and Punjab provinces of Pakistan. *Food Add. Contam. B* **2011**, *4*, 282–288. [CrossRef] [PubMed]
6. Iqbal, S.Z.; Asi, M.R.; Selamat, J. Aflatoxin M1 in milk from urban and rural farmhouses of Punjab, Pakistan. *Food Add. Contam. B* **2014**, *7*, 17–20. [CrossRef] [PubMed]
7. Hussain, I. Micro-Analysis of Aflatoxin M_1 in Dairy Products at Trace Levels and Its Elimination. Ph.D. Thesis, Institute of Chemistry, University of the Punjab, Punjab, Pakistan, 2009.
8. Younus, M.; Abbas, T.; Rafique, K.; Sajid, M.; Aslam, M.; Zafar, M. *Analyses of Selected Heavy Metals and Aflatoxin M1 in Milk for Human Consumption in Jhang City, Pakistan*; Working Paper 12; International Food Policy Research Institute (IFPRI): Washington DC, USA, 2013.
9. Sadia, A.; Jabbar, M.A.; Deng, Y.; Hussain, E.A.; Riffat, S.; Naveed, S.; Arif, M. A survey of aflatoxin M_1 in milk and sweets of Punjab, Pakistan. *Food Control.* **2012**, *26*, 235–240. [CrossRef]
10. Jawaid, S.; Talpur, F.T.; Afridi, H.I. Contamination profile of aflatoxin M_1 residues in milk supply chain of Sindh Pakistan. *Toxicol. Rep.* **2015**, *2*, 1418–1422. [CrossRef] [PubMed]

11. Government of Pakistan. *Economic Survey 2016–2017*; Ministry of Finance, Government of Pakistan: Islamabad, Pakistan, 2017; p. 194.
12. Aziz, S.; Noor-ul-Ain, W.; Majeed, R.; Khan, M.A.; Qayum, I.; Ahmed, I.; Hosain, K. Growth centile charts (anthropometric measurement) of Pakistani pediatric population. *J. Pakistan Med. Assoc.* **2012**, *62*, 367–377.
13. Imtiaz, N.; Yunus, A.W. Comparison of some ELISA kits for aflatoxin M_1 quantification. *J. AOAC Int.* **2019**, *101*. [CrossRef]
14. Barham, G.S.; Khaskheli, M.; Soomro, A.H.; Nizamani, A.A. Screening of adulterants and extent of added water I market milk at central zone of Sindh, Pakistan. *IOSR J. Environ. Sci. Toxicol. Food Technol.* **2014**, *8*, 77–83. [CrossRef]
15. James, C.S. Determination of the fat content of dairy products by the Gerber method. In *Analytical Chemistry of Foods*; Blackie Academic and Professionals: Glasgow, UK, 1995; pp. 93–95.
16. Ling, T.R. *Textbook of Dairy Chemistry*, 3rd ed.; Chapman and Hall: London, UK, 1963; Volume 2.
17. Sharma, R.; Rajput, Y.S. *Detection of Adulterants in Milk—A Laboratory Manual*; Karnal National Dairy Research Institute: Karnal, India, 2012.

toxins

MDPI

Review

Mycotoxins and Mycotoxin Producing Fungi in Pollen: Review

Aleksandar Ž. Kostić [1,*], Danijel D. Milinčić [1], Tanja S. Petrović [2], Vesna S. Krnjaja [3], Sladjana P. Stanojević [1], Miroljub B. Barać [1], Živoslav Lj. Tešić [4] and Mirjana B. Pešić [1]

1 Chemistry and Biochemistry, Faculty of Agriculture, University of Belgrade, Nemanjina 6, 11080 Belgrade, Serbia; danijel.milincic@agrif.bg.ac.rs (D.D.M.); sladjas@agrif.bg.ac.rs (S.P.S.); baracm@agrif.bg.ac.rs (M.B.B.); mpesic@agrif.bg.ac.rs (M.B.P.)

2 Preservation and Fermentation, Faculty of Agriculture, University of Belgrade, Nemanjina 6, 11080 Belgrade, Serbia; tpetrovic@agrif.bg.ac.rs

3 Institute for Animal Husbandry, Autoput 16, 11080 Belgrade, Serbia; vesnakrnjaja.izs@gmail.com

4 Analytical Chemistry, Faculty of Chemistry, University of Belgrade, Studentski Trg 12-16, 11158 Belgrade, Serbia; ztesic@chem.bg.ac.rs

* Correspondence: akostic@agrif.bg.ac.rs

Received: 28 December 2018; Accepted: 21 January 2019; Published: 24 January 2019

Abstract: Due to its divergent chemical composition and good nutritional properties, pollen is not only important as a potential food supplement but also as a good substrate for the development of different microorganisms. Among such microorganisms, toxigenic fungi are extremely dangerous as they can synthesize mycotoxins as a part of their metabolic pathways. Furthermore, favorable conditions that enable the synthesis of mycotoxins (adequate temperature, relative humidity, pH, and a_w values) are found frequently during pollen collection and/or production process. Internationally, several different mycotoxins have been identified in pollen samples, with a noted predominance of aflatoxins, ochratoxins, fumonisins, zearalenone, deoxynivalenol, and T-2 toxin. Mycotoxins are, generally speaking, extremely harmful for humans and other mammals. Current EU legislation contains guidelines on the permissible content of this group of compounds, but without information pertaining to the content of mycotoxins in pollen. Currently only aflatoxins have been researched and discussed in the literature in regard to proposed limits. Therefore, the aim of this review is to give information about the presence of different mycotoxins in pollen samples collected all around the world, to propose possible aflatoxin contamination pathways, and to emphasize the importance of a regular mycotoxicological analysis of pollen. Furthermore, a suggestion is made regarding the legal regulation of pollen as a food supplement and the proposed tolerable limits for other mycotoxins.

Keywords: pollen; fungi; mycotoxins; aflatoxins; ochratoxins; fumonisins; T-2 toxin; zearalenone; deoxynivalenol

Key Contribution: This review gives an overview of scientific data about pollen contamination with different mycotoxins and mycotoxin producing fungi. Also; importance of standard mycotoxicological pollen analysis is emphasized. Inclusion of pollen in the legal regulation; as potential food supplement; is suggested.

---

1. Introduction

Pollen grain, as a male gametophyte of flowering plants, is produced and released from anthers during pollination [1]. Two of the most important pollinators are insects (in the case of entomophilous plants it is, above all, the honey bee (*Apis mellifera* L.)) and, in the case of anemophilous plants, wind. Pollen is prime food for bees due to its amazing diversity of nutritionally important constituents-proteins, lipids, carbohydrates, vitamins, and minerals [2,3]. For the same reasons, floral or bee-collected pollen is potentially a good food supplement for human nutrition [4–8]. Because of great its sensitivity, pollen grain contains a significant quantity of secondary plant metabolites, as part of the plant's defense mechanism, such as different phenolic compounds [9–16] or carotenoids [17,18] and possesses substantial antioxidant properties, which is important for its application as a food supplement [19,20]. Besides the nutritionally important and desirable components, pollen can contain some contaminants such as toxic elements [2,21–23]. Due to optimal water (moisture) content, water activity (a_w), and pH-value, pollen often presents an ideal medium for the development of different microorganisms—bacteria, mold, and yeast. As a result of the presence of mold and yeast, the production of mycotoxins can occur. Mycotoxins are secondary metabolites of different fungi species which are toxic to vertebrates and can lead to some disorders and diseases, or, at worst, death in humans and other animals [24]. The scientific "history" of mycotoxins started in 1962 during a great veterinary crisis when about 100,000 turkeys died in England due to being fed with contaminated peanuts that contained secondary metabolites of *Aspergillus flavus* [24]. The occurrence of mycotoxins in different types of feed and food has been recorded [25–31] and it was found to be strongly dependent on several factors such as climatic conditions (including geographical position of growing region, temperature, and relative humidity) before, during, or after feed/food production [32]. The European Commission (EC Commission Regulation No 1881/2006) sets maximum tolerable levels for several types of mycotoxins (aflatoxins B, G, and/or M, ochratoxin A (OTA), patulin, fumonisins B_1 and B_2, deoxynivalenol, and zearalenone) in different types of foods (nuts, cereals, dried fruits, juices, milk, etc.) [33] but without information pertaining to bee products such as honey, pollen, or bee bread.

The aim of this review is to make a cross-check of current data about contamination of pollen with different types of mycotoxins as well as mycotoxin producing fungi. Also, the effort to emphasize the importance of mycotoxin estimation of pollen samples as obligatory part of their microbiological analysis will be made.

2. Mycotoxins in Pollen

More than a hundred mycotoxins are known, and most of them are produced by some of the species belonging to one of three fungi genera: *Aspergillus*, *Penicillium* and/or *Fusarium* [34]. According to the available literature [35–50] the presence of the following mycotoxins in pollen has been investigated or proved with appropriate analytical methods and analysis: Aflatoxins (AFs), ochratoxins (OTs), fumonisins (FBs), zearalenone (ZEN), deoxynivalenol (DON), and its acetoxy derivative, T-2 toxin (T-2), HT-2 toxin, fusarenon-X, diacetoxyscirpenol, nivalenol, neosolaniol, roridin A, verrucarrin A, α-β-dehydrocurvularin, phomalactone,6-(1-propenyl)-3,4,5,6-tetrahydro-5-hydroxy-4H-pyran-2-one, 5-[1-(1hydroxibut-2-enyl)]-dihydrofuran-2-one and 5-[1-(1-hydroxibut-2-enyl)]-furan-2-one.

2.1. *Aflatoxins*

Aflatoxins are the product of the metabolism of different fungi species which belong to *Aspergillus* genus with *A. flavus* and *A. parasiticus* strains as the main producers [24]. They can be synthetized in fungi's spores and mycelium or secreted as exotoxins [25]. The most toxic and dangerous aflatoxins are aflatoxin B_1 and B_2 (Figure 1) [34]. Both aflatoxin B_1 and B_2 are carcinogenic for humans and animals, and are listed in Group 1 of carcinogenic substances according to International Agency for Research on Cancer (IARC) [51]. The liver is the organ that suffers most from the effects of aflatoxins [52]. Ingestion of these toxins can lead to aflatoxicosis, as an acute form of poisoning, or, in the case of long-term exposure, to the development of liver cancer [52]. Hydroxylated AFB-forms presented in milk are aflatoxin M1 and M2 [24] which are possibly carcinogenic for humans (IARC Group 2A of carcinogenic substances) [34,51]. Furthermore, two other forms of AF exist: Aflatoxin G1 and G2 (Figure 1).

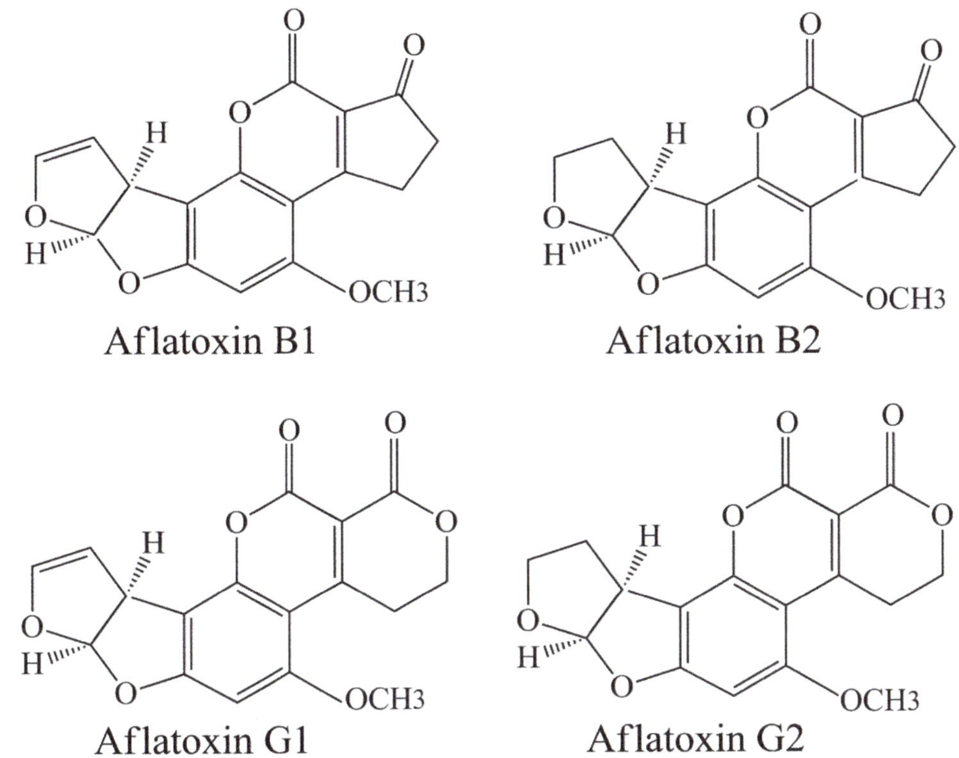

Figure 1. Chemical structures of aflatoxin B_1, B_2, G_1, and G_2.

2.1.1. Contamination of Pollen with Aflatoxins—Possible Ways

Pollen often presents a suitable substrate for the proliferation of various microorganisms due to its favorable moisture content, water activity (a_w), and pH-value. External conditions such as relative humidity and temperature, different stages of pollen production, and storage conditions have been shown to lead to microbiological contamination of pollen [35]. According to data found in the literature, pH-value ranging between 4.0 and 6.5 have been shown to be suitable for the development of bacteria, mold, and yeast while the minimal a_w-values sufficient for the growth of *Aspergillus* and *Penicillium* spp. have been shown to be 0.71 to 0.96 [53] i.e., 0.55 in the case of pollen [54]. Microbiological contamination is strongly pH and temperature dependent and is also conditioned by the type of microorganism [53]. If proper conditions have been achieved in any phase of pollen production, the growth of microbes will occur which can cause aflatoxin production and the contamination of pollen. In addition to production process and human hygiene practices, which are the most important sources of aflatoxin contamination, sometimes microbe growth can be triggered by infected flowering plants [25,48]. Namely, during the flowering and the pollination process, *Aspergillus* spp. spores can germinate on female flower parts. Following this, the toxigenic fungal spores placed in the pollen tubes will grow and further infect the egg-cells [25]. If bees visit these flowers, the contaminated pollen grains will be transferred into the hives. Since there is intensive contact between bees when in the hive (due to highly organized bee societies) their "home" is the third possible source of aflatoxin pollen contamination [48]. As aflatoxins show detrimental effects on bee health, the incidence of these compounds in hives is undesirable. It is for this reason that the occurrence and production of propolis in hives is an effective way for bees to deal with AFs toxicity [55,56] which could indicate that this source of pollen contamination with aflatoxins is at least probable. In the past, aflatoxin occurrence in feed and food was a characteristic of tropic or sub-tropic regions due to favorable climatic conditions. Recently, with climatic changes, which extensively influences weather conditions in temperate areas (such as the majority of Europe), the presence of aflatoxins in these areas is becoming more frequent. The detection of aflatoxins in samples of pollen from the most diverse parts of the world (Table 1) is in accordance with this fact and is becoming a growing problem. Interestingly, in our previous investigation [48] the majority of examined pollen samples were sterile but all were contaminated with AFB_1. This situation confirms three hypotheses:

- There are different ways of pollen contamination with aflatoxin(s).
- These toxins remain in samples with or without presence of appropriate fungi.
- It is extremely important to always perform mycotoxicological analysis together with microbiological characterization of pollen.

Table 1. Toxigenic fungi and concentration level reported for aflatoxins in pollen samples from different countries.

No. of Examined Pollen Samples	Geographical Origin	Analytical Methods	Isolated Mycotoxins Producing Fungi Species	AF Types and Concentration Range(s)	Reference
20	Spain	ELISA test	/	Total AFs: below 5 μg/kg	[35]
20	Spain	HPLC (with fluorescent detection)	/	AFB_1 and AFB_2: below limit detection (BLD)	[37]
87 + 3	Spain + Argentina	HPLC (with fluorescent detection)	*A. flavus* *A. parasiticus*	AFB_1, AFB_2, AFG_1 and AFG_2: not determined.	[38]
5	China	Cyclic voltametry	/	AFB_1: 0.00–0.52 μg/kg	[39,42]
1	Epirus (Western Greece)	HPLC (with fluorescent detection)	not detected	AFB_1: not detected	[40]
45	Slovakia	ELISA test	*A. flavus*, *A. parasiticus.*	Total AFs: 13.60–16.20 μg/kg (in poppy pollen) 3.15–5.40 μg/kg (in rape pollen) 1.20–3.40 μg/kg (in sunflower pollen)	[41]
33	Serbia	ELISA test	*A. flavus*	AFB_1: 3.49–14.02 μg/kg	[44]
20	China	LC-MS/MS	/	AFB_1, AFB_2, AFG_1 and AFG_2: below limit detection (BLD)	[45]
27	Brazil	Qualitative analysis	*A. flavus*	AFB_1 and AFB_2: not determined	[47]
26	Serbia	ELISA test	*A. flavus*	AFB_1: 3.15–17.32 μg/kg	[48]
30	Egypt	Thin-layer chromatography	*A. flavus*	AFB_1 AFB_2, AFG_1 and AFG_2 were not determined.	[49]
9	Portugal	ELISA test	Not detected	Not detected AFB1	[50]

ELISA—enzyme linked immunosorbent assays; AFs—aflatoxins; AFB_1—aflatoxin B_1; AFB_2—aflatoxin B_2; AFG_1—aflatoxin G_1; AFG_2—aflatoxin G_2.

2.1.2. Quantification of Aflatoxins in Pollen Samples

Results of different studies about the determination of aflatoxin content in pollen samples with diverse palynological (botanical) and geographical origins are given in Table 1.

2.2. Ochratoxins

Ochratoxins (OTs) are a group of chemical compounds (Figure 2) derived from shikimic acid metabolic pathway with ochratoxin A (OTA) as a major food contaminant [57]. The main OTs-producers are different *Aspergillus* species with a special emphasis on *Aspergillus niger* strains since they are industrially important due to their applications for enzyme and citric acid production. Furthermore, one species (*P. verrucosum*) belonging to *Penicillium* genus can be the source of ochratoxins [24]. OTA belongs to the IARC 2B group which means that it is a possible carcinogen for humans [51]. The kidneys are the most vulnerable organs effected by OTA. OTA has been noted as having a strong influence on the endemic disease 'Balkan nephropathy', as well as porcine nephropathy, which has been documented in several Scandinavian countries [24].

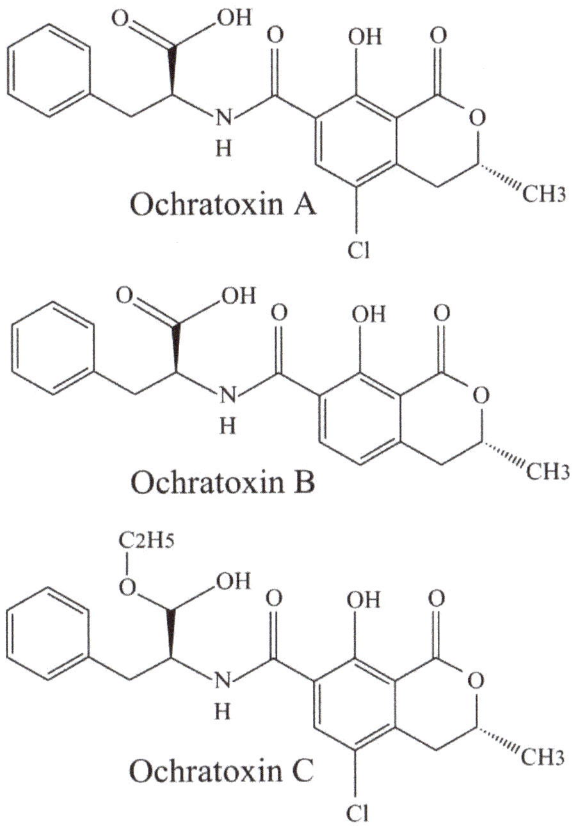

Figure 2. Chemical structures of ochratoxins A, B, and C.

Ochratoxins in Pollen

Besides many types of food (nuts, meat products, barley, oats, rye, wheat, wine, dried fruits, coffee, and coffee products) where the presence of OTA has been recorded [24,57], in some herbs, bottled water [57], and pollen samples, this mycotoxin has also been observed. Xue et al. [45] conducted an examination of 20 bee pollen samples from North China for the presence of OTA p by LC-MS/MS analysis. The obtained results showed that none of the studied pollen samples were contaminated with OTA. These results can be associated with the dry weather conditions during the collection period. The same situation was observed in the case of 20 bee pollen samples that originated from Spain [37]. However, HPLC analysis of 90 Spanish and Argentinian bee pollen samples in [38] confirmed the presence of several *Aspergillus* (*A. carbonarius*, *A. ochraceus* and *A. niger*), and *Penicillium* (*P. verrucosum*) species with the ability to produce OTA. Significant contamination of bee pollen was determined in a case of Slovakian samples [41]. In total, 45 samples were divided in three groups of 15 samples originating from poppy, rape, and sunflower plants. Determined OTA concentration ranges in poppy, rape, and sunflower pollen samples were 6.12 to 10.98 μg/kg, 3.24 to 9.87 μg/kg, and 0.23 to 6.93 μg/kg, respectively. In Spain, by analyzing the toxigenic potential of *A. ochraceus* in various substrates (bee pollen, maize, wheat, and rice) Medina et al. [36] found that OTA production in bee pollen was statistically significantly higher than that found in the production of tested cereals, regardless of the incubation time (7, 14, 21, 28 days). Likewise, positive correlations have been found between the proportion of bee pollen added to the yeast extract sucrose broth inoculated with spores of *A. ochraceus* and OTA level [36]. Based on all of the above, it can be assumed that bee pollen may represent a significant risk factor for the occurrence of OTA in the food chain.

2.3. The Other Mycotoxins Examined in Pollen

2.3.1. Fumonisins

Fumonisins (FBs) are a group of mycotoxins predominantly connected with maize (grown as endophyte in both vegetative or reproductive tissues) and maize products but can be found in many cereals and products made from these plants [24,58]. Although maize is an anemophilic plant due to its high pollen production [7] it is not a rare that bees collect its pollen during the pollen collection season [4]. In that sense, it is possible to find pollen samples contaminated with FBs. The first report about FB food contamination dates back to 1988. The main representative of this mycotoxin group is fumonisin B_1 (FB_1) [24,58]. It is sorted in IARC 2B group of carcinogenic substances [51]. Moreover, fumonisins B_2, B_3, and B_4 also exist (Figure 3) [57]. Fungi belonging to *Fusarium* genus are the most important FBs producers, especially two species: *F. proliferatum* and *F. verticillioides* as well as *A. alternata* from *Alternaria* spp. It is important to point out that the presence of these microbes does not mean that FBs contamination is guaranteed [24]. In an investigation by Kačaniová et al. [41] the presence of both, *F. proliferatum* and *F. verticillioides* was confirmed in thirty i.e., forty-five bee pollen samples, respectively but FBs were quantified only in the samples originating from sunflower (fifteen samples). This observation confirms the previously mentioned hypothesis, that despite the presence of *Fusarium* spp. in some material, appropriate weather conditions or insect damage are necessary for FBs production [24]. The range of FBs concentrations in these samples is given in Table 2.

Table 2. Concentration level reported for mycotoxins other than aflatoxins in pollen samples from different countries.

No. of Contaminated/ Examined Pollen Samples	Geographical Origin	Analytical Methods	Isolated Mycotoxin Producing Fungi Specie(s)	Mycotoxin Types and Concentration Range(s)	Reference
15/45 were contaminated	Slovakia	ELISA test	*F. proliferatum*, *A. alternata* Keissl.	Total FBs: 6.30–12.60 μg/kg	[41]
45	Slovakia	ELISA test	*F. graminearum*	ZEN: 311.00–361.30 μg/kg (in poppy pollen) 137.10–181.60 μg/kg (in rape pollen) 115.60–147.40 μg/kg (in sunflower pollen)	[41]
45	Slovakia	ELISA test	*F. graminearum*, *F. oxysporum*, *F. proliferatum*, *F. sporotrichioides*, *F. verticillioides*	T-2 toxin: 113.90–299.60 μg/kg (in poppy pollen) 197.10-265.70 μg/kg (in rape pollen) 173.60–364.90 μg/kg (in sunflower pollen)	[41]
45	Slovakia	ELISA test	*F. graminearum*, *F. oxysporum*, *F. proliferatum*, *F. sporotrichioides*, *F. verticillioides*	DON: 183.10–273.90 μg/kg (in poppy pollen) 189.60–244.70 μg/kg (in rape pollen) 133.30–203.50 μg/kg (in sunflower pollen)	[41]
2/15	Spain	GC/MS	/	neosolaniol: 22 i.e., 30 μg/kg nivalenol: 1 μg/kg	[43]

ELISA—enzyme linked immunosorbent assays; FBs—fumonisins; ZEN—zearalenone; DON—deoxynivalenol.

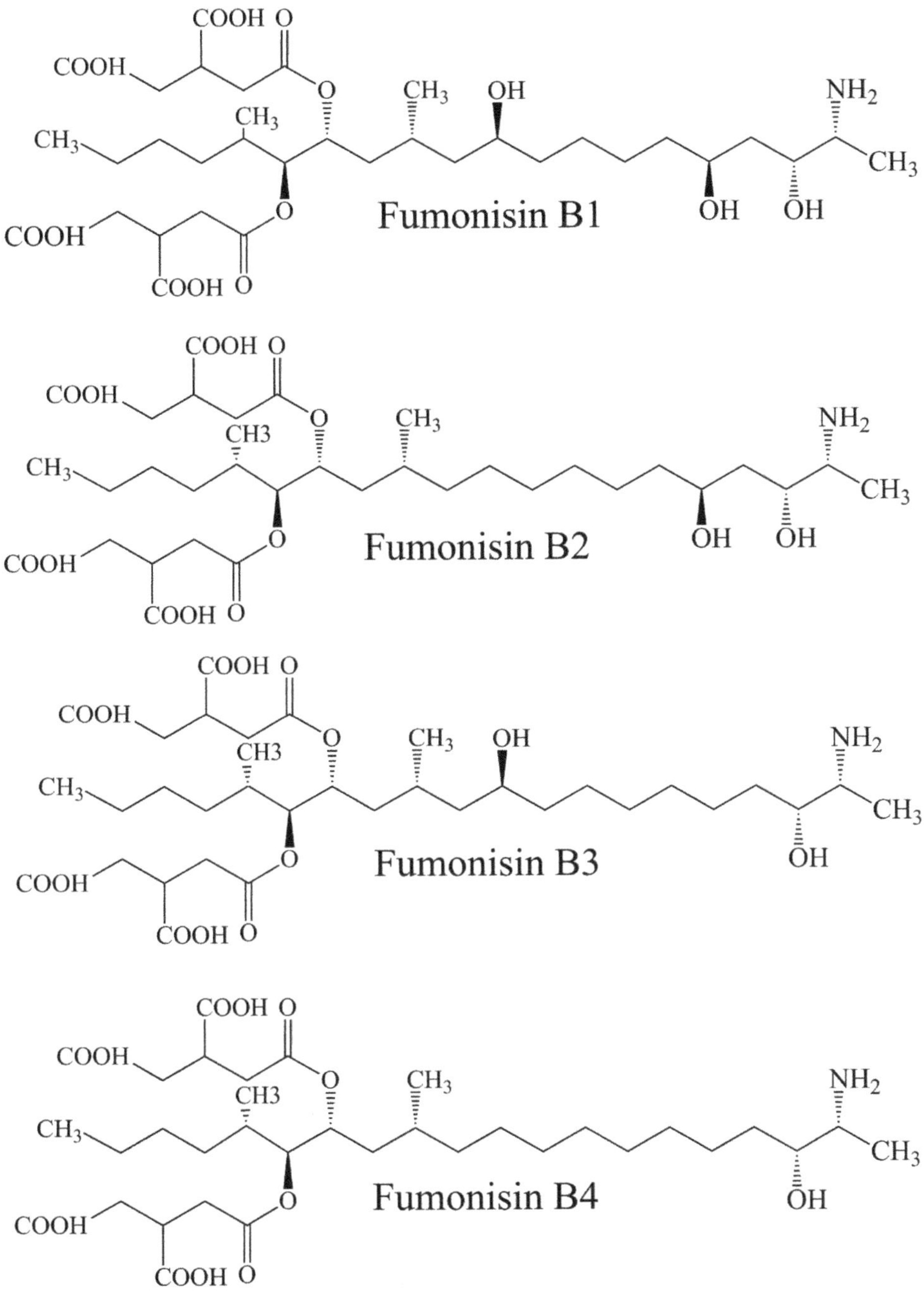

Figure 3. Chemical structures of fumonisins B_1, B_2, B_3, and B_4.

2.3.2. Zearalenone

Zearalenone (ZEN) (Figure 4) is mycoestrogen with limited toxicity that is produced by several *Fusarium* species: *F. graminearum*, *F. culmorum*, *F. crookwellense*, and *F. equiseti*. It is regularly present in crops and crop products [24]. According to IARC this macrocyclic lactone is classified in group 3 which means that it is not classifiable as to its carcinogenicity to humans [51]. In the case of pollen, the significant contamination with ZEN was recorded in Slovakian bee samples [41] (Table 2).

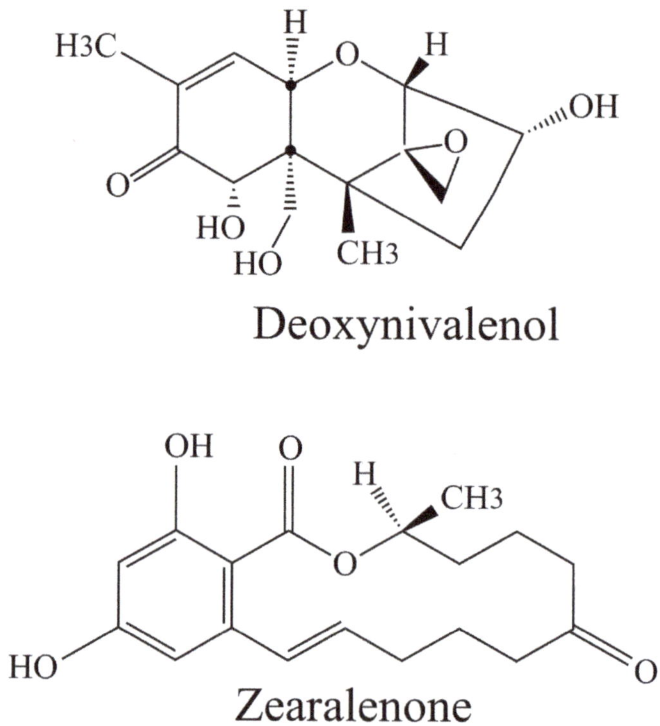

Figure 4. Chemical structures of deoxynivalenol and zearalenone.

2.3.3. Trichothecenes Group of Mycotoxins

In a study from Slovakia [41], the authors also reported the contamination of all examined bee pollen samples with T-2 toxin and deoxynivalenol (Figure 4). Both toxins belong to trichothecene compounds, the sesquiterpenoid metabolites obtained after microbiological activity of several fungi from the following genera: *Fusarium* (primary source), *Trichoderma*, *Myrothecium*, *Phomopsis*, etc., [24]. Together with ZEN, they were the most dominant quantified mycotoxins in the pollen samples. Additionally, the presence of DON and T-2 toxin was checked in fifteen pollen samples from Spain, but the content of these mycotoxins was below limit detection of applied GC/MS method [43]. In the same study, the authors examined the presence of several other *Fusarium* spp. producing mycotoxins: 3-acetyl-deoxynivalenol, fusarenon-X, diacetoxiscirpenol, nivalenol, neosolaniol, and HT-2 toxin. All the above-mentioned compounds belong to trichothecene terpenoid's derivatives. It was determined that some of the samples were contaminated with neosolaniol and nivalenol (Table 2), while all other examined toxins were below limit detection. A report made by Cirigiliano et al. [46] should also be mentioned as their study was the first to detect seven specific mycotoxins (roridin A, verrucarrin A, α-β-dehydrocurvularin, phomalactone,6-(1-propenyl)-3-,4,5,6-tetrahydro-5-hydroxy-4H-pirane-2-one, 5-[1-(1-hydroxibut-2-enyl)]-dihydrofuran-2-one and 5-[1-(1-hydroxibut-2-enyl)]-furan-2-one) in beehives from Argentina with pronounced antifungal effect. Roridin A, verrucarin A, and α-β-dehydrocurvularin were isolated from strains of fungi *Myrothecium verrucaria* while other mycotoxins were obtained as result of *Nigrospora sphaerica* strains activity. Their structures were confirmed by 1D and 2D-NMR spectroscopy.

3. Mycotoxin Producing Fungi in Pollen

The microbiological quality of pollen is equally important as its chemical composition due to its safety use. Although the examination of mycotoxins in pollen began mostly in the last decade, the determination of different microbes (bacteria, mold, and yeast) present in pollen samples started much earlier—at the end of 1970s with studies by Gilliam [59,60]. Considering that a long period of time usually passes between collection of pollen samples and its application as food supplement (or as medicament), there is a great chance for the development of some toxigenic fungi [41]. Their presence may indicate mycotoxin production in pollen with or without their quantification. In that sense, this review also gives information on pollen investigations concerning the presence of mycotoxin producing fungi [41] made without further mycotoxicological analysis. The results of a cross-check of the available literature data, with appropriate comments and information, are given in Table 3.

Table 3. Toxigenic fungi and yeast reported in pollen samples from different countries.

No. of Examined Pollen Samples	Geographical Origin	Detected Microbial Class	Microbial Species or/and Total Microbial	Microbial Count	Observations	Reference
Unknown number of samples of floral and bee-collected almond pollen	unknown	Mold		No. of fungal isolates:	*Mucor* spp. was the dominant mold in floral pollen but not identified in bee-collected pollen. *Aureobasidium pullulans*, *P. corylophilum*, *P. crustosum* and *Rhizopus nigricans* were identified only in bee-collected pollen.	[61]
			Alternaria spp.	6		
			Cladosporium spp.	5		
			Penicillium spp.	5		
			Aspergillus spp.	3		
			Mucor spp.	19		
90 samples of bee pollen	Spain (87 samples) Argentina (3 samples)	Mold	*Aspergillus* section *Nigri*	$1.4 \times 10–2.3 \times 10^2$ cfu/g	The results show the occurrence of different mold species in pollen samples. *Penicillium*, *Alternaria*, and *Aspergillus* spp. were present in 90%, 86.6%, and 80% of samples, respectively. Predominant *Aspergillus* species was *A. niger*. The species of the genus *Fusarium* were isolated in 53.3%.	[38]
			A.flavus +*A. parasiticus*	$1.7 \times 10–2.5 \times 10$ cfu/g		
			Other *Aspergillus* spp.	2×10 cfu/g		
			P. verrucosum	1.4×10^2 cfu/g		
			Other *Penicillium* spp.	$1.3 \times 10^2–4.3 \times 10^3$ cfu/g		
			Fusarium spp.	$16–9.5 \times 10^1$ cfu/g		
			Cladosporium spp.	$6 \times 10–1.4 \times 10^3$ cfu/g		
			Alternaria spp.	$6 \times 10–5.2 \times 10^2$ cfu/g		
			Rhizopus spp.	$2 \times 10–9 \times 10$ cfu/g		
			Mucor spp.	$8–2.2 \times 10^2$ cfu/g		
			Botrytis spp.	$8–3 \times 10$ cfu/g		
			Epicoccum spp.	5–10 cfu/g		
		Yeast	Not specified	$3.6 \times 10^2–7.3 \times 10^3$ cfu/g		
42 samples of dehydrated bee pollen	Brazil	Mold/Yeast	Not specified	Total mold and yeast count: $10^2–1.3 \times 10^4$ cfu/g	About 12% of pollen samples were contaminated with mold and yeast above the limit (1×10^4) for a total mold and yeast proposed by Brazilian legislation.	[62]
30 samples of bee pollen	Slovakia	Microscopic fungi (mold)	*Alternaria* spp. *Cladosporium* spp. *Penicillium* spp. *Fusarium* spp. *Aspergillus* spp. (*A. flavus*, *A. ochraceus*) *Mucor* spp. *Trichoderma* spp. *Acremonium* spp. *Scopulariopsis* spp. *Rhizopus* spp. *Botrytis* spp.	Total mold and yeast count: $1.1 \times 10^2–4.57 \times 10^5$ cfu/g	The dominant fungi isolated from pollen samples were colonies of *A. alternata*, *Cladosporium cladosporoides*, and *Penicillium* spp. Also, the presence of well-known mycotoxicogenic species such as *A. flavus* and *A. ochraceus* were detected.	[63]

Table 3. *Cont.*

No. of Examined Pollen Samples	Geographical Origin	Detected Microbial Class	Microbial Species or/and Total Microbial	Microbial Count	Observations	Reference
19 samples of bee pollen	Mexico	Fungi (mold)	*A. flavus*	Incidence of mold genus (%):	Fungi contamination was generally low. The highest contamination was in three samples handled without packages.	[64]
			Alternaria spp.	3.6%		
			Penicillium spp.	2.9%		
			Fusarium spp.	2.9%		
			Aspergillus spp.	3.6%		
			Mucor spp.	3.1%		
			Rhizopus spp.	0.7%		
8 samples of bee pollen	Slovakia	Mold	*Alternaria* spp. *Cladosporium* spp. *Penicillium* spp. *Aspergillus* spp. *Mucor* spp. *Aureobasidium* spp. *Humicola* spp. *Monodictys* spp. *Paecilomyces* spp. *Rhizopus* spp. *Mortierella* spp. *Trichosporiella* spp. *Harpografium* spp. *Mortierella* spp.	Total mold and yeast count: 107–4688 cfu/g	The results show that in all analyzed samples of pollen 21 fungal species of 13 genera of microscopic fungi were detected. The dominant identified species, over 62% of the isolates belonged to following genera: *Mucor, Rhizopus, Aspergillus, Alternaria*, and *Paecilomyces.*	[65]
28 samples (fresh and dried bee pollen)	Cuba	Mold/Yeast	Not specified	Total mold and yeast count: 10^4–1.5×10^5 cfu/g	All samples had quantified number of mold and yeast above proposed limits (10^4 cfu/g for the fresh and 10^2 cfu/g for dried pollen). Nevertheless, in the dry pollen, a smaller number of high contaminated samples were recorded. Drying could not be used as reliable method for obtaining pollen with acceptable microbiological quality.	[66]
8 samples of commercial bee pollen	Portugal (4 samples) Spain (3 samples) Unknown origin (1 sample)	Mold Yeast	Not specified Individually identified yeast	Total mold and yeast count: <10 to 9.4×10^2 cfu/g	All samples were contaminated with yeast and mold. Further, yeast species were identified, and results indicated the presence of five different genus of yeast which can influence the risk of food-borne illness and spoilage or can serve as an indicator of a lack of hygiene standards.	[67]

Table 3. *Cont.*

No. of Examined Pollen Samples	Geographical Origin	Detected Microbial Class	Microbial Species or/and Total Microbial	Microbial Count	Observations	Reference
Unknown	Portugal	Mold/Yeast	Not specified	Total mold and yeast count: $<10^4$ cfu/g	Generally, yeast and mold were identified in 60% of all examined samples. pH and a_w values had a strong impact on the total microbe number in pollen.	[54]
22 samples of organic bee pollen	Portugal	Mold/Yeast	Not specified	Total mold and yeast count: <10–3560 cfu/g	In all samples of organic bee pollen, the presence of mold and yeast was detected, but their individual species were not identified.	[68]
3 samples of pollen	Algeria	Mold/Yeast	Not specified	Total mold and yeast count: 5×10^4–4×10^5 cfu/g	/	[69]
33 samples of bee pollen	Serbia	Mold	*Alternaria* spp. *Mucor* spp. *Rhizopus* spp. *Cladosporium* spp. *Epicoccum* spp. *Acremonium* spp.	Total mold count: 1×10^3–1×10^5 cfu/g	See Table 1.	[44]
27 samples of dried bee pollen	Brazil	Mold		Total mold count: 1×10^2–5×10^2 cfu/g Incidence of mold genus (%):	Total mold count depends on growing media.	[47]
			Aspergillus spp. (*A. flavus; A. fumigatus; A. versicolor; A. ochraceus; A. carbonarius; A. terreus; A. oryzae*)	85%		
			Cladosporium spp.	63%		
			Penicillium spp. (*P. citrinum; P. citreonigrum; P. glabrum; P. oxalicum*)	41%		
			Alternaria spp.	19%		
			Wallemia spp. and *Eurotium* spp.	11%		
			Mucor spp.	7%		
			Curvularia spp., *Paecilomyces* spp. and *Fusarium* spp. (*F. camptoceras*)	4%		
45 samples of dehydrated bee pollen	Brazil	Mold Yeast	Not specified Identified different species	Total mold and yeast count: <10–7.67×10^3 cfu/g	/	[70]
21 samples of bee pollen (*Melipona* bees)	Brazil	Mold/Yeast	Not specified	/	All samples were sterile without presence of any mold or yeast species.	[71]

Table 3. *Cont.*

No. of Examined Pollen Samples	Geographical Origin	Detected Microbial Class	Microbial Species or/and Total Microbial	Microbial Count	Observations	Reference
40 samples of bee pollen	Italy	Mold	*Cladosporium* spp. *Alternaria* spp. *Humicola* spp. *Mucoraceae* *Acremonium* spp. *Penicillium* spp. (*P. chrysogenum; P. brevicompacticum*) *Aspergillus* spp. (*A. flavus; A. nidulans; A.miger; A. terreus*)	Total mold count: 4–568 cfu/g	In all pollen samples at least one fungal isolate was detected. *Cladosporium* spp. was the most frequently detected mold. *Aspergillus* spp. and *Penicillium* spp., as a potentially mycotoxicogenic mold, were also identified in 8 i.e., 22 pollen samples.	[72]
Dehydrated (electric oven, EO) or lyophilized (L) bee pollen samples	Brazil	Mold/Yeast	Not specified	Total mold and yeast count: 99–242 cfu/g (EO) 16–935 cfu/g (L)	Number of quantified mold and yeast depended on time (April or September) of collection.	[73]
26 samples of bee pollen	Serbia	Mold		Total mold count:	See Table 1	[48]
			Alternaria spp.	1×10^3 cfu/g		
			Mucor spp.	1×10^3 cfu/g		
			Rhizopus spp.	1×10^3 cfu/g		
			Trichoderma spp.	1×10^4 cfu/g		
1 sample of bee pollen	Not known	Mold/Yeast	Not specified	Total mold and yeast count: >21 cfu/g	Presence of yeast and mold can be responsible for the potential presence of toxins in the samples.	[74]
18 samples of commercial bee pollen	Argentina	Mold/Yeast	Not specified	Total mold and yeast count: $<10^2$ cfu/g	The total fungi number is specified for 28% of the samples.	[75]
62 samples of dehydrated bee pollen	Brazil	Mold/Yeast	Not specified	Total mold and yeast count: 1.9×10^2–7.62×10^2 cfu/g	The microbial contamination is dependent on geographical origin of samples.	[76]
8 samples of commercial bee pollen	Algeria	Mold/Yeast	Not specified	Total mold and yeast count: 10^4–2.8×10^5 cfu/g	/	[77]

Table 3. *Cont.*

No. of Examined Pollen Samples	Geographical Origin	Detected Microbial Class	Microbial Species or/and Total Microbial	Microbial Count	Observations	Reference
32 (13 fresh (F) and 19 dried (D) samples of bee pollen)	Bulgaria	Mold	Identified mold: *Aspergillus* spp. *Fusarium* spp. *Penicillium* spp. (*P. brevicompactum*) *Alternaria* spp. *Cladosporium* spp. Other species	Total mold count: 5.6×10^2 –3.7×10^4 cfu/g (F) 150–1.1×10^4 cfu/g (D)	The results show that the values for fungal colony count were significantly lower in the dried pollen samples. 136 fungal isolates were identified. Among detected isolates, genus *Penicillium* was dominant while the genus *Fusarium* was the least fungal contaminant. Dominant species isolated from 14 different samples was *P. brevicompactum*.	[78]
19 samples of stored pollen of five stingless bee species	Brazil	Mold/Yeast	Not specified	Total mold and yeast count: 4.2 $\times 10^1$ cfu/g (1 sample only)	The results show that only for the stored pollen of the stingless bee specie *Frieseomellite varies* it was possible to enumerate mold and yeast.	[79]
bee pollen samples	Colombia	Mold/Yeast	Not specified	Total mold and yeast count: 3×10^2–2×10^5 cfu/g	Number of quantified microbes is strongly dependent on applied temperature for drying of samples.	[80]

4. Legislations of Mycotoxins Level in Food and Pollen

In order to prevent undesirable consequences and to protect consumers health, the European Commission, as well as some other international agencies, have proposed maximum permissible concentrations (MPC) for several mycotoxins in different types of food [33,81]. Maximum permissible concentrations vary due to differences in food origin and greater/less possibility of contamination with mycotoxins, as well as because of smaller or larger intake in meals. For instance, the MPC for AFB_1 alters from 0 to 8 μg/kg [33]. Zero tolerance is established for milk and dairy products due to regular daily consumption while the maximal value has been proposed for groundnut-based food. Furthermore, for sensitive groups (such as infants and children), special lower limits have been usually established. The proposed limits are subject to corrections as a result of the development of new, more precise, and sensitive analytical methods for determining the content of mycotoxins [81]. In Table 4 current EU MPC values for some food types are given.

Table 4. Examples for the current maximum permissible concentrations (MPC) for some mycotoxins in different types of food/food supplements.

Food/Food Supplements	Mycotoxin(s)	MPC Value(s)	Reference
Groundnuts used as components for food production	AFB_1	8 μg/kg	[33]
	Sum of AFB_1, AFB_2, AFG_1 and AFG_2	15 μg/kg	
Groundnuts for direct human consumption	AFB_1	2 μg/kg	[33]
	Sum of AFB_1, AFB_2, AFG_1 and AFG_2	4 μg/kg	
Dried fruits used as components for food production	AFB_1	5 μg/kg	[33]
	Sum of AFB_1, AFB_2, AFG_1 and AFG_2	10 μg/kg	
Dried fruits for direct human consumption	AFB_1	2 μg/kg	[33]
	Sum of AFB_1, AFB_2, AFG_1 and AFG_2	4 μg/kg	
Raw milk used for consumption and dairy productions, infant formulae and infant-milk	AFB_1	0 μg/kg	[33]
	Sum of AFB_1, AFB_2, AFG_1 and AFG_2	0 μg/kg	
Unprocessed cereals	OTA	5 μg/kg	[33]
Cereals based products	OTA	3 μg/kg	[33]
Instant coffee	OTA	10 μg/kg	[33]
Roasted coffee	OTA	5 μg/kg	[33]

The Scientific Committee of Food requested and obtained from the European Food Safety Authority (EFSA) current data for Tolerable Weekly Intake (TWI) for OTA—0.12 μg/kg of body weight (bw) [82]. Recently, EFSA published new information about the potential increase of maximum allowable level (from 4 to 10 μg/kg) for total AFs in peanuts and processed products, requested by EU Commission [83]. The CONTAM panel (EFSA Panel on Contaminants in the Food Chain) strongly opposed this request due to the significant increase of cancer risk (factor value = 1.6–1.8). For other mycotoxins proposed Tolerable Daily Intake (TDI) values are: 2 μg/kg bw for nivalenol, 0.25 μg/kg bw for ZEN [84], 2 μg/kg (provisional maximum TDI) for FBs [85], 1 μg/kg bw for DON [86], 0.1 μg/kg bw for the sum of T-2 and HT-2 toxins [87], 0.06 μg/kg for combined trichothecenes mycotoxins group [33]. In these legislations, there is no information about proposed limits for mycotoxins in pollen. In 2008 Campos et al. [2] proposed that in the case of AFB_1 occurrence in pollen the MPC value should be set at 2 μg/kg i.e., 4.2 μg/kg for total AFs. To the best of our knowledge, this is the only proposal which defines the level of some mycotoxins in pollen. Since this paper gives an overview about the presence of different mycotoxins in pollen samples originating from various locations around the world, it will be of great importance to define some tolerable levels for other fungi-produced toxins in pollen, especially for OTA. Moreover, current values for AFB_1 and AFs should be reconsidered and checked due to an increasingly frequent aflatoxin contamination caused by climatic changes. Special concerns exist due to mixed (cross) contamination of pollen samples as confirmed by the presented data. Previously, several authors [32,88,89] confirmed that some combined mycotoxins have a more distinct detrimental effect on human health. Furthermore, Manafi et al. [90] have shown that AFs and T-2 toxin synergistically influenced the decrease of total serum protein and albumin levels in broiler chickens as well as decreased antibody titers. It is therefore of the utmost importance to evaluate the toxicological impact of mycotoxin combinations on animal and human health risks.

5. Conclusions and Future Perspectives

Pollen could be used as a food supplement which can be attributed to its appropriate chemical composition. The microbiological quality of pollen is equally important as its nutritional characteristics. The fungal contamination of different feed/food, including pollen will be more frequent as a result of intensive climatic changes. The quality of pollen can be significantly influenced by the presence of toxigenic fungi. Since it has been proved that the absence of microbial contamination in pollen does not exclude the presence of mycotoxins, mycotoxicological analyses should also be included as a regular control measure together with microbiological tests. Since aflatoxins and ochratoxins are proven as carcinogenic substances, their presence in pollen is extremely undesirable. Therefore, it is important to monitor mold and mycotoxin levels in feed/food in order to avoid adverse health effects. The incorporation of pollen as a food supplement in current legislation will be useful. Proposed quality parameters need to cover tolerable daily/weekly intake for different mycotoxins as well as their sum. In order to obtain reliable and accurate recommendations for pollen quality control, further studies on the toxicological impact of mycotoxin combinations should be conducted.

Author Contributions: All authors participated in the creation and conceptualization of the article. A.Ž.K. and D.D.M. conducted the literature search. A.Ž.K., D.D.M., and M.B.P. wrote the manuscript. T.S.P., V.S.K., S.P.S., M.B.B., Ž.L.T. and M.B.P. controlled and critically reviewed language and manuscript content during preparation. All authors read and approved final manuscript.

Funding: Ministry of Education, Science and Technological Development of the Republic of Serbia, projects TR31069, OI172017, 46010 and TR31023.

Acknowledgments: Authors would like to thank to Vladimir Kostić for technical support in Graphical apstract preparation.

Conflicts of Interest: All authors declare no conflict of interest.

References

1. Borg, M.; Brownfield, L.; Twell, D. Male gametophyte development: A molecular perspective. *J. Exp. Bot.* **2009**, *60*, 1465–1478. [CrossRef]
2. Campos, G.R.M.; Bogdanov, S.; Almeida-Muradian, L.B.; Szczesna, T.; Mancebo, Y.; Frigerio, C.; Ferreira, F. Pollen composition and standardization of analytical methods. *J. Apic. Res.* **2008**, *47*, 154–161. [CrossRef]
3. Bogdanov, S. Pollen: Collection, harvest, composition, quality. In *Bee Product Science (The Pollen Book)*; 2012; Chapter 1; Available online: http://www.bee-hexagon.net/pollen/collection-harvest-composition-quality/ (accessed on 23 January 2019).
4. Kostić, A.Ž.; Barać, M.B.; Stanojević, S.P.; Milojković-Opsenica, D.M.; Tešić, Ž.L.; Šikoparija, B.; Radišić, P.; Prentović, M.; Pešić, M.B. Physicochemical properties and techno-functional properties of bee pollen collected in Serbia. *LWT Food Sci. Technol.* **2015**, *62*, 301–309. [CrossRef]
5. Kostić, A.Ž.; Kaluđerović, L.M.; Dojčinović, B.P.; Barać, M.B.; Babić, V.B.; Mačukanović-Jocić, M.P. Preliminary investigation of mineral content of pollen collected from different Serbian maize hybrids—Is there any potential nutritional value? *J. Sci. Food Agric.* **2017**, *97*, 2803–2809. [CrossRef] [PubMed]

6. Kostić, A.Ž.; Pešić, M.B.; Trbović, D.; Petronijević, R.; Dramićanin, A.; Milojković-Opsenica, D.M.; Tešić, Ž.L. Fatty acid's profile of Serbian bee-collected pollen—Chemotaxonomic and nutritional approach. *J. Apic. Res.* **2017**, *56*, 533–542. [CrossRef]
7. Kostić, A.Ž.; Mačukanović-Jocić, M.P.; Špirović Trifunović, B.D.; Vukašinović, I.Ž.; Pavlović, V.B.; Pešić, M.B. Fatty acids of maize pollen-quantification, nutritional and morphological evaluation. *J. Cereal Sci.* **2017**, *77*, 180–185. [CrossRef]
8. Conte, P.; Del Caro, A.; Balestra, F.; Piga, A.; Fadda, C. Bee pollen as a functional ingredient in gluten-free bread: A physical-chemical, technological and sensory approach. *LWT Food Sci. Technol.* **2018**, *90*, 1–7. [CrossRef]
9. Campos, M.; Markham, M.R.; Mitchell, K.A.; da Cuhna, A.P. An approach to the characterization of bee pollens via their flavonoid/phenolic profiles. *Phytochem Anal.* **1997**, *8*, 181–185. [CrossRef]
10. Serra-Bonvehí, J.; Torrentó, S.M.; Lorente, C.E. Evaluation of polyphenolic and flavonoid compounds in honeybee-collected pollen produced in Spain. *J. Agric. Food Chem.* **2001**, *49*, 1848–1853. [CrossRef]
11. Campos, M.G.; Webby, F.B.; Markham, M.R.; Mitchell, K.A.; da Cuhna, A.P. Age-induced diminution of free radical scavenging capacity in bee pollens and the contribution of constituent flavonoids. *J. Agric. Food Chem.* **2003**, *51*, 742–745. [CrossRef]
12. Di Paola-Naranjo, R.D.; Sánchez, S.J.; Paramás, A.M.G.; Gonzalo, J.C.R. Liquid chromatographic-mass spectrometric analysis of anthocyanin composition of dark blue bee pollen from *Echium Plantagineum*. *J. Chromatogr. A* **2004**, *1054*, 205–210. [CrossRef] [PubMed]
13. Almaraz Abarca, N.; Campos da Graça, M.; Ávila-Reyes, J.A.; Naranjo-Jiménez, N.; Corral, J.H.; González-Valdez, L.S. Antioxidant activity of polyphenolic extract of monofloral honeybee-collected pollen from mesquite (*Prosopis juliflora, Leguminosae*). *J. Food Compos. Anal.* **2007**, *20*, 119–124. [CrossRef]
14. Ferreres, F.; Pereira, D.M.; Valentão, P.; Andrade, P.B. First report of noncoloured flavonoids in *Echium plantagineum* bee pollen: Differentation of ismomers by liquid chromatography/ion trap mass spectometry. *Rapid Commun. Mass Spectrom.* **2010**, *24*, 801–806. [CrossRef]
15. Ares, A.M.; Valverde, S.; Bernal, J.L.; Nozal, M.J.; Bernal, J. Extraction and determination of bioactive compounds from bee pollen. *J. Pharm. Biomed. Anal.* **2018**, *147*, 110–124. [CrossRef] [PubMed]
16. De-Melo, A.A.M.; Estevinho, L.M.; Moreira, M.M.; Delerue-Matos, C.; da Silva de Freitas, A.; Barth, O.M.; de Almeida-Muradian, L.B. Phenolic profile by HPLC-MS, biological potential, and nutritonal value of a promising food: Monofloral bee pollen. *J. Food Biochem.* **2018**, *42*, e12536. [CrossRef]
17. Almeida-Muradian, L.B.; Pamplona, L.C.; Coimbra, S.; Ortrud, M.B. Chemical composition and botanical evaluation of dried bee pollen pellets. *J. Food Compos. Anal.* **2005**, *18*, 105–111. [CrossRef]
18. Mărgăoan, R.; Mărghitaş, L.A.; Dezmirean, D.S.; Dulf, F.V.; Bunea, A.; Socaci, S.A.; Bobiş, O. Predominant and secondary pollen botanical origins influence the carotenoid and fatty acid profile in fresh honeybee-collected pollen. *J. Agric. Food Chem.* **2014**, *62*, 6306–6316. [CrossRef]
19. Krystyjan, M.; Gumul, D.; Ziobro, R.; Korus, A. The fortification of biscuits with bee pollen and its effect on physicochemical and antioxidant properties in biscuits. *LWT Food Sci. Technol.* **2015**, *63*, 640–646. [CrossRef]
20. De Florio Almeida, J.; Soares dos Reis, A.; Serafini Heldt, L.F.; Pereira, D.; Bianchin, M.; de Moura, C.; Plata-Oviedo, M.V.; Haminiuk, C.W.I.; Ribeiro, I.S.; Fernades Pinto da Luz, C.; et al. Lyophilized bee pollen extract: A natural antioxidant source to prevent lipid oxidation in refrigerated sausages. *LWT Food Sci. Technol.* **2017**, *76*, 299–305. [CrossRef]
21. Kostić, A.Ž.; Pešić, M.B.; Mosić, M.D.; Dojčinović, B.P.; Natić, M.N.; Trifković, J.Đ. Mineral content of some bee-collected pollen from Serbia. *Arch. Ind. Hyg. Toxicol.* **2015**, *66*, 251–258. [CrossRef]
22. Sattler, J.A.G.; de Melo Machado, A.A.; do Nascimento, K.S.; de Melo Pereira, I.L.; Mancini-Filho, J.; Sattler, A.; de Almeida-Muradian, L.B. Essential minerals and inorganic contaminants (barium, cadmium, lithium, lead and vanadium) in dried bee pollen produced in Rio Grande do Sul State, Brazil. *Food Sci. Technol.* **2016**, *36*, 505–509. [CrossRef]
23. Altunaltmaz, S.S.; Tarhan, D.; Aksu, F.; Barutçu, U.B.; Or, M.E. Mineral element and heavy metal (cadmium, lead and arsenic) levels of bee pollen in Turkey. *Food Sci. Technol.* **2017**, *37* (Suppl. S1), 136–141. [CrossRef]
24. Bennet, J.W.; Klich, M. Mycotoxins. *Clin. Microbiol. Rev.* **2003**, *16*, 497–516. [CrossRef]
25. Hanssen, E.; Jung, M. Control of aflatoxins in the food industry. *Pure Appl. Chem.* **1973**, *35*, 239–250. [CrossRef]

26. Bosco, F.; Mollea, C. Mycotoxins in food. In *Food Industrial Processes—Methods and Equipment*; Valdez, B., Ed.; Intech Open Limited: London, UK, 2012; Chapter 10; pp. 169–200. ISBN 978-953-307-905-9.
27. Stanković, S.; Lević, J.; Ivanović, D.; Krnjaja, V.; Stanković, G.; Tančić, S. Fumonisin B1 and its co-occurrence with other fusariotoxins in naturally-contaminated wheat grain. *Food Control* **2012**, *23*, 384–388. [CrossRef]
28. Krnjaja, V.; Mandić, V.; Lević, J.; Stanković, S.; Petrović, T.; Vasić, T.; Obradović, A. Influence of N-fertilization on Fusarium head blight and mycotoxin levels in winter wheat. *Crop Prot.* **2015**, *67*, 251–256. [CrossRef]
29. Abrunhosa, L.; Morales, H.; Soares, C.; Calado, T.; Vila-Cha, A.S.; Pereira, M.; Venâncio, A. A review of mycotoxins in food and feed products in Portugal and estimation of probable daily intake. *Crit. Rev. Food Sci. Nutr.* **2016**, *56*, 249–265. [CrossRef]
30. Bijelić, Z.; Krnjaja, V.; Stanković, S.; Muslić-Ružić, D.; Mandić, V.; Škrbić, Z.; Lukić, M. Occurrence of moulds and mycotoxins in grass-legume silages influenced by nitrogen fertilization and phenological phase at harvest. *Rom. Biotech. Lett.* **2017**, *22*, 12907–12914.
31. Krnjaja, V.; Stanković, S.; Obradović, A.; Petrović, T.; Mandić, V.; Bijelić, Z.; Božić, M. Trichothecene genotypes of *Fusarium graminearum* populations isolated from winter wheat crops in Serbia. *Toxins* **2018**, *10*, 460. [CrossRef]
32. Smith, M.-C.; Madec, S.; Coton, E.; Hymery, N. Natural co-occurrence of mycotoxins in foods and feeds and their in vitro combined toxicological effects. *Toxins* **2016**, *8*, 94. [CrossRef] [PubMed]
33. EC Commission. Setting of maximum levels for certain contaminants in foodstuffs—Regulation No. 1881/2006. *Official J. of the EU.* **2006**, *L364*, 5–24.
34. Van Egmond, H.P. Mycotoxins: Risks, regulations and European co-operation. *J. Nat. Sci. Matica Srpska Novi Sad.* **2013**, *125*, 7–20. [CrossRef]
35. Serra-Bonvehi, J.; Escolà Jordà, R. Nutrient composition and microbiological quality of honey bee-collected pollen in Spain. *J. Agric. Food Chem.* **1997**, *45*, 725–732. [CrossRef]
36. Medina, Á.; González, G.; Sáez, J.M.; Mateo, R.; Jiménez, M. Bee pollen, a substrate that stimulates ochratoxin A production by *Aspergillus ochraceus* Wilh. *Syst. Appl. Microbiol.* **2004**, *27*, 261–267. [CrossRef] [PubMed]
37. Garcia-Villanova, R.J.; Cordón, C.; González-Paramás, A.M.; Aparicio, P.; Garcia Rosales, M.E. Simultaneous immunoaffinity column cleanup and hplc analysis of aflatoxins and ochratoxin a in spanish bee pollen. *J. Agric. Food Chem.* **2004**, *52*, 7235–7239. [CrossRef] [PubMed]
38. González, G.; Hinojo, M.J.; Mateo, R.; Medina, A.; Jiménez, M. Occurrence of mycotoxin producing fungi in bee pollen. *Int. J. Food Microbiol.* **2005**, *105*, 1–9. [CrossRef] [PubMed]
39. Zaijun, L.; Zhongyun, W.; Xiulan, S.; Yinjun, F.; Peipei, C. A sensitive and highly stable electrochemical impendace immunosensor based on the formation of silica gel-ionic liquid biocompatible film on the glassy carbon electrode for the determination of aflatoxin B1 in bee pollen. *Talanta* **2010**, *80*, 1632–1637. [CrossRef]
40. Pitta, M.; Markaki, P. Study of aflatoxin B1 production by *Aspergillus parasiticus* in bee pollen of Greek origin. *Mycotoxin Res.* **2010**, *26*, 229–234. [CrossRef]
41. Kačaniová, M.; Juráček, M.; Chlebo, R.; Kňazovická, V.; Kadasi-Horáková, M.; Kunová, S.; Lejková, J.; Haščik, P.; Mareček, J.; Šimko, M. Mycobiota and mycotoxins in bee pollen collected from different areas of Slovakia. *J. Environ. Sci. Health Part B* **2011**, *46*, 623–629. [CrossRef]
42. Vidal, J.C.; Bonel, L.; Ezquerra, A.; Hernández, S.; Bertolín, J.R.; Cubel, C.; Castillo, J.R. Electrochemical affinity biosensors for detection of mycotoxins: A review. *Biosens. Bioelectron.* **2013**, *49*, 146–158. [CrossRef]
43. Rodríguez-Carasco, Y.; Font, G.; Mañes, J.; Berrada, H. Determination of mycotoxins in bee pollen by gas chromatography–tandem mass spectrometry. *J. Agric. Food Chem.* **2013**, *61*, 1999–2005. [CrossRef] [PubMed]
44. Petrović, T.; Nedić, N.; Paunović, D.; Rajić, J.; Matović, K.; Radulović, Z.; Krnjaja, V. Natural mycobiota and aflatoxin B1 presence in bee pollen collected in Serbia. *Biotechnol. Anim. Husb.* **2014**, *30*, 731–741. [CrossRef]
45. Xue, X.; Selvaraj, J.N.; Zhao, L.; Dong, H.; Liu, F.; Liu, Y.; Li, Y. Simultaneous determination of aflatoxins and ochratoxin a in bee pollen by low-temperature fat precipitation and immunoaffinity column cleanup coupled with LC-MS/MS. *Food Anal. Methods* **2014**, *7*, 690–696. [CrossRef]
46. Cirigliano, A.M.; Rodríguez, M.A.; Godeas, A.M.; Cabrera, G.M. Mycotoxins from beehive pollen mycoflora. *J. Sci. Res. Rep.* **2014**, *3*, 966–972. [CrossRef]
47. Valadares Deveza, M.; Keller, K.M.; Affonso Lorenzon, M.C.; Teixeira Nunes, L.M.; Oliveira Sales, E.; Barth, O.M. Mycotoxicological and palynological profiles of commercial brands of dried bee pollen. *Braz. J. Microbiol.* **2015**, *46*, 1171–1176. [CrossRef]

48. Kostić, A.Ž.; Petrović, T.S.; Krnjaja, V.S.; Nedić, N.M.; Tešić, Ž.L.; Milojković-Opsenica, D.M.; Barać, M.B.; Stanojević, S.P.; Pešić, M.B. Mold/aflatoxin contamination of honey bee collected pollen from different Serbian regions. *J. Apic. Res.* **2017**, *56*, 13–20. [CrossRef]
49. Hosny, A.S.; Sabbah, F.M.; El-Bazza, Z.E. Studies on microbial decontamination of Egyptian bee pollen by γ-irradiation. *Egypt Pharm. J.* **2018**, *17*, 190–200. [CrossRef]
50. Estevinho, L.M.; Dias, T.; Anjos, O. Influence of the storage conditions (frozen vs dried) in health-related lipid indexes and antioxidants of bee pollen. *Eur. J. Lipid Sci. Technol.* **2018**, *2018*, 1800393. [CrossRef]
51. Vidal, A.; Mengelers, M.; Yang, S.; De Saeger, S.; De Boevre, M. Mycotoxin biomarkers of exposure: A comprehensive review. *Compr. Rev. Food Sci. Food Saf.* **2018**, *17*, 1127–1155. [CrossRef]
52. Neal, G.E. Genetic implications in the metabolism and toxicity of mycotoxins. *Toxicol. Lett.* **1995**, *82/83*, 861–867. [CrossRef]
53. Magan, N.; Lacey, J. Effect of temperature and pH on water realtions of field and storage fungi. *Trans. Br. Mycol. Soc.* **1984**, *82*, 71–81. [CrossRef]
54. Estevinho, L.M.; Rodrigues, S.; Pereira, A.P.; Feás, X. Portugese bee pollen: Palynological study, nutritional and microbiological evaluation. *Int. J. Food Sci. Technol.* **2012**, *47*, 429–435. [CrossRef]
55. Niu, G.; Johnson, R.M.; Berenbaum, M.R. Toxicity of mycotoxins to honeybees and its amelioration by propolis. *Apidologie* **2011**, *42*, 79–87. [CrossRef]
56. Temiz, A.; Şener Mumcu, A.; Özkök Tüylü, A.; Sorkun, K.; Salih, B. Antifungal activity of propolis samples collected from different geographical regions of Turkey against two food-related molds, *Aspergillus versicolor* and *Penicillium aurantiogriseum*. *Gida* **2013**, *38*, 135–142. [CrossRef]
57. Tao, Y.; Xie, S.; Xu, F.; Liu, A.; Wang, Y.; Chen, D.; Pan, Y.; Huang, L.; Peng, D.; Wang, X.; et al. Ochratoxin A: Toxicity, oxidative stress and metabolism (Review). *Food Chem. Toxicol.* **2018**, *112*, 320–331. [CrossRef] [PubMed]
58. Cendoya, E.; Chiotta, M.L.; Zachetti, V.; Chulze, S.N.; Ramirez, M.L. Fumonisins and fumonisin-producing *Fusarium* occurrence in wheat and wheat by products: A review. *J. Cereal Sci.* **2018**, *80*, 158–166. [CrossRef]
59. Gilliam, M. Microbiology of pollen and bee bread: The yeasts. *Apidologie* **1979**, *10*, 43–53. [CrossRef]
60. Gilliam, M. Microbiology of pollen and bee bread: The genus *Bacillus*. *Apidologie* **1979**, *10*, 269–274. [CrossRef]
61. Gilliam, M.; Prest, D.B.; Lorenz, B.J. Microbiology of pollen and bee bread: Taxonomy and enzimology of molds. *Apidologie* **1989**, *20*, 53–68. [CrossRef]
62. Carelli Barreto, L.M.R.; Cunha Funari, S.R.; de Oliveira Rosi, R. Composição e qualidade do pólen apícola proveniente de sete estados Brasileiros e do distrito federal. *Bol. Ind. Anim.* **2005**, *62*, 167–175.
63. Kačániová, M.; Pavličová, S.; Haščík, P.; Kociubinski, G.; Kňazovická, V.; Sudzina, M.; Sudzinova, J.; Fikselová, M. Microbial communities in bees, pollen and honey from Slovakia. *Acta Microbiol. Immunol. Hung.* **2009**, *56*, 285–295. [CrossRef] [PubMed]
64. Bucio Villalobos, C.M.; López Preciado, G.; Martínez Jaime, O.A.; Torres Morales, J.J. Micoflora asociada a granos de polen recolectados por abejas domésticas (*Apis mellifera* L.). *Rev. Electron. Nova Sci.* **2010**, *4*, 93–103. [CrossRef]
65. Brindza, J.; Gróf, J.; Bacigálová, K.; Ferianc, P.; Tóth, D. Pollen microbial colonization and food safety. *Acta Chim. Slov.* **2010**, *3*, 95–102.
66. Puig-Peña, Y.; del-Risco-Ríos, C.A.; Álvarez-Rivera, V.P.; Leiva-Castillo, V.; García-Neninger, R. Comparación de la calidad microbiológica del polen apícola fresco y después de un proceso de secado. *Rev. CENIC. Cienc. Biol.* **2012**, *43*, 23–27.
67. Nogueira, C.; Iglesias, A.; Feás, X.; Estevinho, M.L. Commercial bee pollen with different geographical origins: A comprehensive approach. *Int. J. Mol. Sci.* **2012**, *13*, 11173–11187. [CrossRef] [PubMed]
68. Feás, X.; Pilar Vázquez-Tato, M.; Estevinho, L.; Seijas, J.A.; Iglesias, A. Organic bee pollen: Botanical origin, nutritional value, bioactive compounds, antioxidant activity and microbiological quality. *Molecules* **2012**, *17*, 8359–8377. [CrossRef] [PubMed]
69. Hani, B.; Dalila, B.; Saliha, D.; Daoud, H.; Mouloud, G.; Seddik, K. Microbiological sanitary aspects of pollen. *Adv. Environ. Biol.* **2012**, *6*, 1415–1420.
70. De-Melo Machado, A.A.; Estevinho, M.L.M.F.; Almeida-Muradian, L.B. A diagnosis of the microbiological quality of dehydrated bee-pollen produced in Brazil. *Lett. Appl. Microbiol.* **2015**, *61*, 477–483. [CrossRef]
71. Santa Bárbara, M.; Machado, C.S.; da Silva Sodré, G.; Dias, L.G.; Estevinho, L.M.; Lopes de Carvalho, C.A. Microbiological assessment, nutritional characterization and phenolic compounds of bee pollen from *Mellipona mandacaia* Smith, 1983. *Molecules* **2015**, *20*, 12525–12544. [CrossRef]

72. Nardoni, S.; D'Ascenzi, C.; Rocchigiani, G.; Moretti, V.; Mancianti, F. Occurrence of molds from bee pollen in Central Italy—A preliminary study. *Ann. Agric. Environ. Med.* **2016**, *23*, 103–105. [CrossRef]
73. De-Melo Machado, A.A.; Fernandes Estevinho, M.L.M.; Gasparotto Sattler, J.A.; Rodrigues Souza, B.; da Silva Freitas, A.; Barth, O.M.; Bicudo Almeida-Muradian, L. Effect of processing conditions on characteristics of dehydrated bee-pollen and correlation between quality parameters. *LWT Food Sci. Technol.* **2016**, *65*, 808–815. [CrossRef]
74. Grabowski, N.T.; Klein, G. Microbiology of processed edible insect products—Results of a preliminary survey. *Int. J. Food Microbiol.* **2017**, *243*, 103–107. [CrossRef] [PubMed]
75. Libonatti, C.; Andersen-Puchuri, L.; Tabera, A.; Varela, S.; Passucci, J.; Basualdo, M. Caracterización microbiológica de polen comercial. Reporte preliminar. *Rev. Electron. Vet.* **2017**, *18*, 1–5.
76. Aparecida Soares de Arruda, V.; Vieria dos Santos, A.; Figueiredo Sampaio, D.; da Silva Araújo, E.; de Castro Peixoto, A.L.; Fernandes Estevinho, L.M.; de Almeida-Muradian, B.L. Microbiological quality and physicochemical characterization of Brazilian bee pollen. *J. Apic. Res.* **2017**, *56*, 231–238. [CrossRef]
77. Adjlane, N.; Hadj Ali, L.M.; Benamara, M.; Bounadi, O.; Haddad, N. Qualite microbiologique du pollen produit par les apiculteurs et commercialise en Algerie. *Rev. Microbiol. Ind. San et Environ.* **2017**, *11*, 31–39.
78. Beev, G.; Stratev, D.; Vashin, I.; Pavlov, D.; Dinkov, D. Quality assessment of bee pollen: A cross sectional survey in Bulgaria. *J. Food Qual. Hazards Control* **2018**, *5*, 11–16. [CrossRef]
79. Figueredo Santa Bárbara, M.; Santiago Machado, C.; da Silva Sodré, G.; de Lima Silva, F.; Alfredo Lopes de Carvalho, C. Microbiological and physicochemical characterization of the pollen stored by stingless bees. *Braz. J. Food Technol.* **2018**, *21*, e2017180. [CrossRef]
80. Zuluaga-Domínguez, C.; Serrato-Bermudez, J.; Quicazán, M. Influence of drying-related operations on microbiological, structural and physicochemical aspects for processing of bee-pollen. *Eng. Agric. Environ. Food.* **2018**, *11*, 57–64. [CrossRef]
81. Arroyo-Manzanares, N.; Huertas-Pérez, J.F.; García-Campaña, A.M.; Gámiz-Gracia, L. Mycotoxin analysis: New proposals for sample treatment. *Adv. Chem.* **2014**, *2014*, 547506. [CrossRef]
82. European Food Safety Authority (EFSA). Opinion of the Scientific Panel on contaminants in the food chain of the EFSA on a request from the Commission related to ochratoxin A in food. *EFSA J.* **2006**, *4*, 365. [CrossRef]
83. EPSA Panel on Contaminants in the Food Chain (Contam); Knutsen, H.K.; Barregard, L.J.A.; Bingami, M.; Bruschweiler, B.; Ceccatelli, S.; Cottrill, B.; Dinovi, M.; Edler, L.; Grasl-Kraup, B.; et al. Effect on public health of a possible increase of the maximum level for 'aflatoxin total' from 4 to 10 μg/kg in peanuts and processed products thereof, intended for direct human consumption or use as an ingredient in foodstuffs-statement. *EFSA J.* **2018**, *16*, 5175. [CrossRef]
84. EPSA Panel on Contaminants in the Food Chain (Contam). Scientific Opinion on the risks for public health related to the presence of zearalenone in food. *EFSA J.* **2011**, *9*, 2197. [CrossRef]
85. EPSA Panel on Contaminants in the Food Chain (Contam); Knutsen, H.K.; Alexander, J.; Barregard, L.J.A.; Bingami, M.; Bruschweiler, B.; Ceccatelli, S.; Cottrill, B.; Dinovi, M.; Grasl-Kraup, B.; et al. Scientific Opinion on the risks for human and animal health related to the presence of modified forms of certain mycotoxins in food and feed. *EFSA J.* **2014**, *12*, 3916. [CrossRef]
86. EPSA Panel on Contaminants in the Food Chain (Contam). Risks to human and animal health related to the presence of deoxynivalenol and its acetylated and modfied forms in food and feed. *EFSA J.* **2017**, *15*, 4718. [CrossRef]
87. EPSA Panel on Contaminants in the Food Chain (Contam). Scientific Opinion on the risks for animal and public health related to the presence of T-2 and HT-2 toxin in food and feed. *EFSA J.* **2011**, *9*, 2481. [CrossRef]
88. Šegvić Klarić, M. Adverse effects of combined mycotoxins. *Arh. Ind. Hyg. Toxikol.* **2012**, *63*, 519–530. [CrossRef]
89. Šegvić Klarić, M.; Rašić, D.; Peraica, M. Deleterious effects of mycotoxin combinations involving Ochratoxin, A. *Toxins* **2013**, *5*, 1965–1987. [CrossRef] [PubMed]
90. Manafi, M.; Umakantha, B.; Mohan, K.; Narayana Swamy, H.D. Synergistic effects of two commonly contaminating mycotoxins (Aflatoxin and T-2 toxin) on biochemical parameters and immune status of broiler chickens. *World Appl. Sci. J.* **2012**, *17*, 364–367.

toxins

MDPI

Review

Multi-(myco)toxins in Malting and Brewing By-Products

Kristina Mastanjević [1,*], Jasmina Lukinac [1], Marko Jukić [1], Bojan Šarkanj [2], Vinko Krstanović [1] and Krešimir Mastanjević [1]

1 Faculty of Food Technology Osijek, Josip Juraj Strossmayer University of Osijek, F. Kuhača 20, 31000 Osijek, Croatia; ptfosptfos2@gmail.com (J.L.); ptfosptfos@gmail.com (M.J.); vkrstano@ptfos.hr (V.K.); kmastanj@gmail.com (K.M.)

2 Department of Food Technology, University North, University Center Koprivnica, Trg dr. Žarka Dolinara 1, 48000 Koprivnica, Croatia; bsarkanj@unin.hr

* Correspondence: kristinahabschied@gmail.com; Tel.: +385-31-224-411

Received: 11 December 2018; Accepted: 4 January 2019; Published: 9 January 2019

Abstract: Fungi, yeasts, and bacteria are common microorganisms on cereals used in malting and brewing industries. These microorganisms are mostly associated with the safety and quality of malt and beer, but also with the health safety of by-products used in animal nutrition. The real problem is their harmful metabolites—toxins that, due to their thermostable properties, can easily be transferred to malting and brewing by-products. Besides fungal metabolites, other toxins originating from plants can be harmful to animal health. Precise and accurate analytical techniques broadened the spectrum of known toxins originating from microorganisms and plants that can pose a threat to animal health. Multi-(myco)toxin analyses are advanced and useful tools for the assessment of product safety, and legislation should follow up and make some important changes to regulate yet unregulated, but highly occurring, microbial and plant toxins in malting and brewing by-products used for animal feed.

Keywords: Multi-toxins; mycotoxins; malting and brewing by-products; animal feed

Key Contribution: The emerging multi-toxins can represent a serious threat to human and animal health. The unknown consequences of these compounds are yet to be pinpointed, but as a precaution measure, the presence of such toxins in malting and brewing by-products should be actively monitored and, with time, included into legislation regarding animal feed.

---

1. Introduction

Beer can be produced not only from barley malt, but other cereals (malted or unmalted wheat, corn, rice, sorghum) can be added to the brew to cut production costs or improve the colloidal stability of beer. During malting and brewing processes, by-products are formed and adjusted to be reused as feed. Nowadays, modern processes can be applied and transform these by-products into functional foods or additives for the food industry. Many studies on myco- and other toxins are available in professional and scientific literature, but only a few investigations have focused on by-products formed during the malting and brewing process, such as germ and rootlets, spent grains, and spent yeast. These by-products represent a nutritious and valuable, low-cost source of feed for livestock. In terms of using it as functional food, malting and brewing (M/B) by-products can be added to different cereal products to enhance their nutritional properties. However, the above-mentioned by-products can also be contaminated with mycotoxins and other toxins originating from plants or other microorganisms. Most mycotoxins are thermo-stable and can survive the unit operations applied during the malting and brewing processes. In addition, M/B by-products are susceptible to microbial contamination and have a short shelf life (due to the original high water content). To ensure the food safety for the final consumers, whether animals or humans, it is necessary to investigate, determine, and regularly monitor the existing toxins in these by-products.

Encouraged by the recent findings of our research on multi-toxins in malting and brewing by-products [1,2], we decided to round up similar papers and point out how important and urgent this matter of emerging myco- and other toxins in animal feed is.

2. The Production of Malting and Brewing By-Products

Malting is a process of forced germination of cereal grains to develop enzymes needed to break down starch and proteins to more simple molecules (glucose, maltose, maltotriose, maltodextrines, and aminoacids), so that yeast can be employed in alcohol production. When germ and rootlets are shown, germination is halted by hot air to dry out the excess water. After that, the germinated grain is subjected to degermination, germ and rootlets get separated, and are adapted to be sold as animal feed. Since they are low in bulk density, they are mixed with barley and malt dust and small barley grains and pelletized [3]. Fungi originating from field and storage facilities thrive under malting conditions.

The brewing process consists of several continuous phases. Firstly, milled malt is subjected to mashing (slow heating with water). The residual precipitate from this process is called spent grains and the liquid part, wort, undergoes further processing to become beer.

After precipitation, wort is boiled, and, at this point, hops are added. When boiling is done, hopped wort is cooled and separated from hop residues/spent hops. Spent hops get discarded, but some of it ends up in the trub/cold break (the protein precipitate leftover after cooling of wort).

After cooling, hopped wort is inoculated with yeast and left for main fermentation in a fermentation vessel. During this time, the yeast biomass increases three- to six fold, and at the end of this phase, the yeast biomass is called spent yeast [4].

All these by-products can be utilized for human or animal consumption. However, the native microflora and accompanying produced or bio-transformed (myco)toxins can be a cause of serious economy- and health-related problems for humans and animals.

3. Common Fungal Species in Malting and Brewing

The most common fungal species in malting and brewing by-products are different Fusarium species. *F. acuminatum*, *F. anthophilium*, *F. avanceum*, *F. cerealis* (crookwellense), *F. chlamydosporum*, *F. culmorum*, *F. equiseti*, *F. graminearum*, *F. heterosporum*, *F. nygamai*, *F. oxysporum*, *F. poae*, *F. proliferatum*, *F. sambucinum*, *F. semitectum*, *F. sporotrichoides*, *F. subglutaminans*, *F. tricintum*, and *F. verticilioides* all belong to genus *Fusarium*, but only some of them show inclination to invade grains in our region. According to Krstanović et al. [5], the most prevalent species in Eastern Croatia is *F. graminearum*, being present in both wheat and barley samples (Figure 1). The highest contaminations (20%) were detected in samples from Nova Gradiška. *F. culmorum* was found in significantly lower percentages, with 3.1% being the highest contamination, and statistically did not represent any threat to human health. Generally, colder Northern European regions support the growth of *F. culmorum*, and warmer parts [6] of Europe fight battles with *F. graminearum*. However, the global climate changes caused a shift in Fusarium species' distribution. Namely, according to Parikka et al. [7], *F. culmorum* lost its place in being the prevalent specie in Nordic countries. *F. graminearum* is expected to populate the Northern European areas very soon. Ward et al. [8] reported on toxicity of *F. graminearum* in North America and similar research was conducted in China, where the results indicated that more aggressive isolates (producing a wider variety of mycotoxins) are taking over [9].

(a) **(b)**

Figure 1. Heavily infected malt (*F. graminearum*) before (**a**) and after (**b**) kilning.

Other fungal species invading cereals and making problems in malting and brewing industries include: *Alternaria* spp., *Epicoccum* spp., *Penicillium* spp., *Aspergillus* spp., etc. In research conducted on Slovakian barley, many fungal species were detected: *Acremonium strictum*, *Alternaria* spp., *Aspergillus niger*, *Aspergillus flavus*, *Botrytis cinerea*, *Cladosporium herbarum*, *Cladosporium cladosporioides*, *Cochliobolus sativus* (*Helminthosporium sativum*), *Epicoccum nigrum*, *Fusarium* spp., *Chrysonilia sitophila*, *Microdochium nivale*, *Mucor* sp., *Nigrospora sphaerica*, *Penicillium* sp., *Phaeosphaeria nodorum* (*Septoria nodorum*), *Pyrenophora teres* (*Helminthosporium teres*), *Rhizoctonia* sp., *Rhizopus nigricans*, *Stemphylium* sp., *Tanatephorus cucumeris* (*Rhizoctonia solani*), *Trichoderma* sp., and *Verticillium albo-atrum* [10].

In a study conducted by Gonzalez Pereyra et al. [11], the toxigenic species belonging to the genera, *Fusarium*, *Aspergillus*, *Penicillium*, *Cladosporium*, *Geotrichum*, and *Alternaria*, were identified in malted barley and brewer's spent grains. The detected *Fusarium* spp. made 30% and *Aspergillus* spp. made 27.3% of contamination, and 30% of brewer's grain samples were contaminated with *Mucorales*. Most prevalent *Fusarium* species in brewer's spent grain were *Fusarium verticillioides* (50%) and *Fusarium proliferatum* (25%) while *Fusarium equiseti* (12.5%) and *Fusarium oxysporum* (12.5%).

Piacentini et al. [12] published the results from their research involving the detection of fungal species and mycotoxins (deoxynivalenol (DON) and fumonisins (FB_1 and FB_2)) on malting barley. They also reported that the most prevalent genera were *Fusarium*. *F. graminearum* and *F. verticillioides* were the dominant species, with an incidence of occurrence of 26% and 12%. Other fungal genera determined in their research included *Penicilium*, *Alternaria*, and *Rhizophus*. Mycotoxin analysis revealed that DON contaminated 18% of samples while FB_1 and FB_2 were detected at a lesser extent, 10% for FB_1 and 1% for FB_2.

Fungi on malting barley are problematic because of the favourable conditions [13], which support fungal mycelium growth and mycotoxins' production. Fungi-thriving process conditions are assisted by the easily available nutrients from the grain, emergence in water, exposure to lower temperatures (10–14 °C) and aeration, high air humidity (>90%), and overall high grain moisture during germination and the first phase of kilning (cca. 45% grain moisture). Subsequent mycotoxin biosynthesis occurs at almost every stage of malting and brewing. Colder temperatures during steeping and germination enable fungi to produce mycotoxins. Mycotoxins synthesis occurs [2,14] during the whole malting process, especially after kilning, and continues during brewing. The fermenting stage also allows mycotoxins' content increase. In case there are some leftover spores, mycotoxins can be found even during aging. Some toxins stem from yeast (tryptophol) and are produced after the inoculation of wort. According to our findings [2], mycotoxin concentrations vary during the brewing stage, and after inoculation, mycotoxins (mainly DON) can show a significant increase in comparison to wort.

4. General Toxicity and Occurrence of Multi- and Myco-Toxins

Fungal infections are not unusual and they often contaminate cereals, diminishing their quality and lowering their marketability. However, the most important problem related to fungi is the occurrence of different myco- and other toxins that, directly or indirectly, affect human and animal health.

The sources of human exposure to mycotoxins can be direct (from plant-derived foods contaminated with toxins) or indirect (through a carryover of mycotoxins and their metabolites in animal products, such as meat and eggs) [15]. The exposure to air and dust containing toxins can also cause serious consequences to human health [16]. Therefore, it is important to wear suitable protection (masks-preferably with HEPA (High Efficiency Particulate Air) filter, goggles) during handling of infected grain [17].

As is well-known, mycotoxins influence vital organs and tissues, having carcinogenic properties, inducing immunosuppression and reproductive problems in animals and humans [15]. Most mycotoxins affect the health of humans and animals by causing nephropathy, infertility, cancer, or death [18]. Intense research on mycotoxins reveals that cca. 400 mycotoxins are known till today [19] and about 200 of them belong to the group of trichothecenes [18]. Especially dangerous are mycotoxins that accumulate in animal tissues, meat, and fat, because they can cause problems when ingested by humans.

Some studies report aflatoxins, fumonisins, trichothecenes, ochratoxin A, and zearalenone as the most significant mycotoxin in the barley-beer chain [20,21], but for animal feed, only seven mycotoxins: Aflatoxin B_1 (AFB_1), DON, zearalenone (ZEA), fumonisins (FUM) and ochratoxin A (OTA), T-2, and HT-2 are regulated by the EU (European Union) legislation [22–24] and are under strict governmental control.

Aflatoxins (AFs) are produced by different species of *Aspergillus* fungi (*A. flavus*, *A. parasitous*, and *A. nomius*). They can occur in different foods (groundnuts, tree nuts, maize, corn, rice, figs, and other dried foods, spices, crude vegetable oils, and cocoa beans) as a result of fungal contamination before and after harvest. Aflatoxins are potentially toxic, carcinogenic, mutagenic, and immunosuppressive compounds [25]. There are 18 different types of AFs, with the major members being AFB_1, AFB_2, AFG_1, AFG_2, AFM_1, and AFM_2. AFB_1 is the only mycotoxin whose maximum content in mg/kg (relative to a feedstuff moisture content of 12%) is determined by the Commission Directive 2003/100/EC [22]. Even though AF's can end up in beer, the European Union has not set a maximum allowable limit for AFB_1 in beer, although there is a limit for barley and malt [26]. AFB_1 has been described as a high carcinogenic natural toxin and has been classified as a Group 1 (carcinogenic to humans) [27]. AFB_1 can be found in beer [28–30] and raw materials, such as barley [31,32] and corn [33,34], used for the production of beer. Thus, AFB_1 can be found in brewing by-products that can be used as animal feed and cause health problems for animals, as confirmed by Gerbaldo et al. [35] and Gonzalez Pereyra et al. [11].

Most *Fusarium* spp. produce trichothecenes, and amongst them, *F. graminearum*, *F. culmorum*, *F. proliferatum*, and *F. equiseti* are characterized as main producers of trichotecenes [36]. However, many other fungal species can produce trichotecenes: *Myrotecium*, *Trichoderma*, *Cephalosporium*, *Verticimonosporium*, and *Stachybotrys* [37]. According to Krstanović et al. [5], the most prevalent species on cereals in Eastern Croatia are *F. graminearum* Schw. (teleomorph *Gibberella zeae*) and *F. culmorum* (Wm. G. Sm.) Sacc. (teleomorph unknown) can be found at a lesser extent. DON, being the most prevalent mycotoxin in cereals, is identified as one of the indicators of the quality and safety of barley and wheat used for malting [38]. In the research by Velić et al. [39] conducted with *F. graminearum*, a correlation between fungal contamination and DON appearance was shown. DON is identified as a protein synthesis inhibitor in eukaryotic cells due to its ability to bind to ribosomes. Low to moderate concentrations of DON can affect the health of animals, causing anorexia (lower economical profit), and higher doses can cause vomiting [40,41]. Besides DON, other trichotecenes can be found in malting and brewing by-products, and many of them occur in coherence with their derives. Many trichotecenes are well described in scientific literature and are well known in popular circles. They are nivalenol (NIV),

T-2 toxin, diacetoxyscirpenol (DAS), etc. However, they have not been yet included into legislative regarding barley, malt, or beer for that matter. This can represent a serious problem since many of them can be found in malting and brewing by-products used as animal feed [1,2].

ZEA is one of the infamous mycotoxins that acts similarly to estrogen and is related with clinical manifestations of various estrogenic effects in humans and farm animals. The exposure to this mycotoxin can been seen in pubertal changes in young children in Puerto Rico [42] and gynecomastia with testicular atrophy in southern Africa [43]. Although ZEA is mostly determined in maize, it can easily be found in other cereals, such as wheat and barley. Maize is one of the unmalted cereals that can be added to a brew to cut costs and improve the colloidal stability of beer, and thus can cause damage to the brewing industry if ZEA is present in the raw material. ZEA has different modified forms that derive from two different metabolic phases:

1. Phase I includes: α-zearalenol, β-zearalenol (α-ZEL and β-ZEL), zearalenone-4-O-β-glucoside (ZEA-14Glc), zearalenone-16-O-β-glucoside (ZEA-16Glc), and zearalenone sulphate (ZEA-14-sulphate, ZEA-14S), and many of them have the ability to transform into their basic form during digestion in the mammalian intestinal tract [44]. ZEA-14Glc can generate complex compounds, such as zearalenone-malonyl-glucoside (ZEA-MalGlc), zearalenone-di-hexoside (ZEA-di-hexoside), and zearalenone-pentosylhexoside [45].
2. Phase II derivate, β-zearalenol-4-glucoside (β-ZEL-4Glc), may be formed by maize itself during xenobiotic detoxification [46] and can easily be transferred back into the original form-zearalenone [47].

Binder et al. [48] conducted research in which they monitored the biotransformation of two plants and one fungal metabolite: ZEN-14-sulfate, ZEN-14-O-β-glucoside, and ZEN-16-O-β-glucoside in pigs. Although the total biological recoveries in urine and faeces were below 50%, the authors attributed this to the extensive metabolization by intestinal bacteria to yet unknown metabolites and offered in conclusion that the complete hydrolyzation of all monitored metabolites (ZEN-14-sulfate, ZEN-14-O-β-glucoside, and ZEN-16-O-β-glucoside) occurred, contributing to the total ZEN toxicity.

Fumonisins are mycotoxins produced by fungi belonging to the genus, *Fusarium*. There are many detected fumonisins, but type B fumonisins are most prevalent in feed and food chains [49]. Fumonisins B_1, B_2, B_3, and B_4 all comprise this group, with FB_1 and FB_2 being the most common members [50,51]. Although primarily found in corn, fumonisins can be found in other cereals. Symptoms of fumonisin intoxication are various [52] for various animals. Pigs, for example, can suffer from porcine pulmonary edema [53] and horses can manifest hemorrhagic-liquefactive brain lesions. In humans, neural tube defects and growth retardation in children can occur [54]. According to Voss and Riley [52], FB_1 is characterized as a potent carcinogen as some reports link exposure to high doses of FB_1 to human esophageal cancers [55]. However, no strong empirical relationship has been established between the manifestation of cancer and FB_1, so FB_1 is still being considered as possibly carcinogenic to humans and is categorized as a group 2B carcinogen [56]. Fumonisins can also exist in modified forms, which Braun and Wink [56] call "cryptic" fumonisins. They form when basic fumonisins undergo hydrolyzation, usually during digestion in the gastrointestinal tract. In 2018, EFSA (European Food Safety Authority) [57] issued a scientific opinion on fumonisins and their modified and hidden forms in feed. According to EFSA, modified forms are extractable from the matrix and can be formed during biotransformations in fungal, plant, or animal organisms or during thermal and chemical processes to which feed and food can be subjected to. The term 'hidden forms' refers to the fraction of fumonisins strongly bonded with non-covalent bonds to the matrix and they are basically non-extractable.

Ochratoxins are secondary metabolites primarily produced by *Penicillium* and *Aspergillus* species. This group involves ochratoxin (OTA), its methyl ester, its ethyl ester also known as ochratoxin C (OTC), 4-hydroxyochratoxin A (4-OH OTA), ochratoxin B (OTB) and its methyl and ethyl esters, and ochratoxin α (OTα) [58]. OTA is nephrotoxic, carcinogenic, teratogenic, genotoxic, and immunotoxic [59]. OTA is one of the highest occurring ochratoxins and is incorporated into legislative recommendations and regulations. As reported by Huff and Doerr [60], OTA and aflatoxin can act synergistically and significantly retard the growth of broiler chickens.

T-2 and HT-2 are *Fusarium* mycotoxins incorporated into EU legislative. Like other mycotoxins, their occurrence is climate dependent and they usually occur together. They are toxic to most animals and humans as they interfere with protein synthesis and haematopoiesis inhibition, lymphoid depletion, and necrotic lesions [61]. Plants and yeasts can modify these mycotoxins into different glucosides: T-2 3-O-glucoside (T-2Glc) and HT-2-O-3-glucoside (HT-2Glc), HT-2-di-glucoside (HT-2di-Glc) [62], di-glucosides (T-2di-Glc, and HT-2di-Glc) [63]. Yeasts of the genus, *Trichomonascus* and *Blastobotrandy*, have three biotransformation levels for the T-2 toxin: Acetylation to 3-acetyl-T-2; glycosylation; and the removal of an isovaleryl group from the molecule, leading to the formation of neosolaniol [64]. Other modified forms of T-2 and HT-2 mycotoxins include: HT-2-malonyl-glucoside, hydroxy-HT-2-glucoside, dehydro-HT-2-glucoside, T2-triol-glucoside, and T-2-feruloyl-T-2 toxin [65].

Emerging mycotoxins in food and feed, such as enniatins (ENNs) and beauvericin (BEA), have been under the scientific light for the past few years. The toxicity of enniantin B (ENNB) and BEA has been recently described by a group of authors in EFSA's [57] external scientific report, where it was reported that ENNB is genotoxic to mice and BEA targeted thyroid, kidneys, and reproductive systems in both sexes. Repeated oral exposure to BEA seems to affect male mice more, but further studies should be conducted.

Besides these mycotoxins and their derivates stemming from plants itself, there are many others that can be found in malting and brewing by-products. Some of them are reported in a recent publication by Mastanjević et al. [1]. Emerging toxins are currently the focal point of the scientific community since these toxins are not included into legislation and some of them are very poorly studied; in a sense, there is little or no information of how they affect human or animal health.

5. Multi- and Myco-Toxins in Malting and Brewing By-Products

Since M/B by-products are rich in proteins, digestible fibre material, vitamins, and minerals, their utilization as feed supplements has become an alternative for animal feed [11,66,67]. These by-products represent low cost and nutritious animal feed, but also promote rumen fermentation and digestion and overall act beneficially on animal health [11].

Cavaglieri et al. [67] confirmed the correlation between fungal contamination and natural incidence of selected mycotoxins in barley rootlets. Namely, Cavaglieri et al. [67] found FB_1 contamination in 100% of barley rootlet samples. However, FB_1 levels found in this substrate were subjectively high (254–2043 μg/kg); while Gerbaldo et al. [35] carried out another study that confirmed AFB_1 contamination on brewer's grain.

In a research conducted by Gonzalez Pereyra et al. [11], the detected mycotoxins in malting barley and BSG (brewer's spent grain) were fumonisin B_1 and AFB_1. Although FB_1 toxin levels were below the recommended values (104–145 μg/kg), the occurrence of the fumonisins' contamination in these samples was 100%. Aflatoxin B_1 was detected in 18% of brewer's spent grain samples (19–45 μg/kg) and was presumably produced during storage [68]. Other mycotoxins (AFB_2, AFG_1, AFG_2, or ZEA) were not found in the analysed samples. This indicates that, in the case of brewer's grain and other barley by-products intended for animals, mycotoxin contamination increases with the storage time if storage conditions are not closely controlled. Similar research was conducted by Habschied et al. [69], in which the ZEA concentration in barley germ and rootlets after degermination increased with prolonged storage time and higher water activity (a_w).

Barley germ and rootlets are the result of the ending phase of the malting process [67]. After the final phase of drying, degermination is carried out. Germ and rootlets have a lower water content, but because of their high nutritive value in amino acids and fat content, they are susceptible to microbial degradation. Inappropriate storage and management of these materials can lead to the loss of nutritive substances and mycotoxins concentration increase [67,69]. Figure 2 shows a significant reduction of germ/rootlets in malt severely contaminated with *Fusarium graminearum* (sample b). The higher the initial *Fusarium* contamination, the poorer the germ development will be.

(a) (b)

Figure 2. Germ/rootlets of (**a**) healthy malt (**b**) severely infected malt with *F. graminearum*.

As mentioned before, Mastanjević et al., Habschied et al., and Krstanović et al. investigated several cases in which mycotoxins and multi-toxins were determined in M/B by-products [1,2,69–71]. They reported that not only mycotoxins can be found in such by-products, but also in many other toxins and originating from plants or other sources.

Brewer's spent grain is obtained after the filtration of mash and are a moist brewing by-product. Spent grain can be used as animal feed in various forms: fresh (wet) (Figure 3), ensiled, or dehydrated (dried) grain [72]. Spent grain is often used as a supplement for lactating cows [73]. Since spent grains and spent yeast are very nutritious, and contain high amounts of available water [11], they are susceptible to microbial growth and are very perishable. Before storage, any excess water must be removed to ensure microbial safety and to prolong the shelf life. Spent yeast also must be treated in additional processes to inactivate the living yeast cells because ruminants have a complex mixture of micro flora to help breakdown the cellulose into simpler carbohydrates they can use for energy. Although small amounts of live yeast can be safe for ruminants, excessive amounts of live yeast cells can disrupt the gut microbiology. This ends up as a gas build up that can block their airways, causing suffocation and even death. Drying also improves the storability of these by-products and enables easier transportation and handling [73].

Figure 3. Spent grains and cold break (protein residue) after filtration.

A research conducted by Mastanjević et al. [2] followed the transition of eight *Fusarium culmorum* mycotoxins (fusarenone-X (FUS-X), 3-ADON, DAS, HT-2, T-2, DON, NIV, ZEA) from starting wheat to beer and associated by-products (steeping water, germ/rootlets, spent grains, and spent yeast). The results showed that steeping water can withhold extremely high DON concentrations amounting up to 20,326 µg/kg. Germ/rootlets also retain high amounts of DON (5636 µg/kg) and NIV (1691 µg/kg). This is in accordance with an earlier research by Krstanović et al. [71] conducted on wheat samples with different initial *F. culmorum* infection levels (0–50%). Namely, germ/rootlets here too retained significantly high DON concentrations (5000–12,600 µg/kg), depending on the infection level. The germ is rich in fats and made of soft tissue, hence it is a place of the most intense fungal growth, which explains why DON levels in germ are so high. Similar research was conducted by Habschied et al. [69]

on barley malt where ZEA was the followed mycotoxin. This research showed that ZEA levels in malt bran and germ samples were significantly higher than in malt flour (endosperm) in almost all growing conditions.

This affirms that it is necessary to monitor and determine the exact types and quantity of mycotoxins in cereals, final products, and by-products to ensure the food and feed safety [70]. In addition, according to Habschied et al. [69], storage conditions for these by-products are of immense importance. Thus, attention should be paid to ensure the suitable storage conditions.

Except mycotoxins, many malting and brewing by-products contain other toxins stemming from the plant itself, residual fungicides, or herbicides applied in the field. Namely, plants possess two detoxifying mechanisms (chemical modification and compartmentalization) that help them transfigure (myco)toxins to a less toxic form [47]. To explain this, Berthiller et al. [47] and Rychlik et al. [74] described deoxynivalenol-3-glucoside (D-3-G) as a product of a detoxifying plant mechanism, where glycosylation of DON occurs via a conjugation reaction and the result is D3G. Modified mycotoxins usually co-occur with free mycotoxins [47], and can even exceed the concentration of free form in processed foods and feed [75,76].

In research by Mastanjević et al. [1], a wide array of multi-toxins was detected in malting and brewing by-products. Most abundantly found mycotoxins in spent grains and spent yeast were lotaustralin and tryptophol. Other detected myco- and multi-toxins were: Aurofusarin, beauvericin, brevianamid F, chrysogin, culmorin, 5-hydroxyculmorin, 15-hydroxyculmorin, deoxynivalenol, deoxynivalenol-3-glucoside, linamarin, tentoxin, and zearalenon. Some of these toxins can be found in Table 1 and Figure 4.

Table 1. Known and emerging detectable toxins that can be found in M/B by-products.

		Mycotoxins	Source
Fungal	*Fusarium* mycotoxins	Deoxynivalenol, Nivalenol, 3-Acetyldeoxynivalenol, 15-Acetyldeoxynivalenol, Zearalenon, HT-2 toxin, T2-toxin, Moniliformin, Chrysogin, Culmorin,, Aurofusarin, Fumonisin B_1, Fumonisin B_2, Fumonisin B_3, Beauvericin, Enniantins (A, A_1, B, B_1)	[1,2,12,15,77]
	Other species mycotoxins (*Aspergillus, Alternaria*)	Aflatoxin B_1, Tentoxin, Brevianamid F	[1,11,77]
Yeast metabolite		Tryptophol	[1]
Plant metabolites		DON-3-glucoside, Lotaustralin, Linamarin	[1,77]

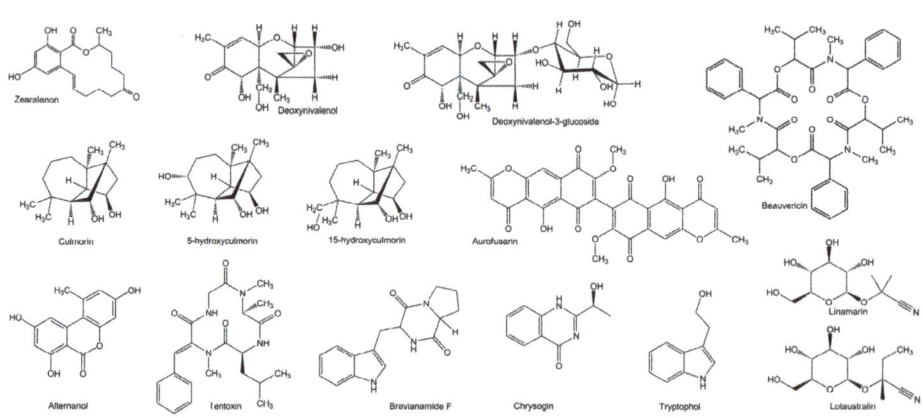

Figure 4. Chemical structure of some of known and emerging toxins.

Since the spectrum of found mycotoxins is very broad, an update regarding the detection and determination of emerging multitoxins in legislation should be made to ensure the health of animals and, subsequently, humans. Surveillance of mycotoxin concentrations should be regular and stricter and analysis spectrum should be expanded, especially on masked mycotoxins (3-acetyl-DON (3-ADON) and 15-acetyl-DON (15-ADON), cyanogenic glucosides (D-3-G), and yeast secondary metabolites (tryptophol).

6. Malting and Brewing By-Products Adding Value to Different Food Products

Besides the use in animal feed, malting and brewing by-products can be used to enrich and improve the functional properties of different food products [4,78,79]. Considering the increasing, almost pandemic, occurrence of obesity, diabetes, cancer, and cardiovascular diseases in today's world, a general idea of functional food derived from food industry by-products has been diligently implemented in contemporary research investigations.

Being produced in large quantities and available all year long, brewing by-products, such as spent grains, spent hops, and spent yeast, can be utilized in food production by adding nutritive value to different products. The main constituents -minerals, nitrogen, and carbon—are the most important compounds from a biotechnological point of view and they have a tremendous potential for use in biotechnological processes. As mentioned before, being rich in carbon and nitrogen makes them suitable for the extraction of sugars, proteins, acids, and antioxidants [4].

Brewer's spent grain and spent yeast can also be applied in the baking industry as a source of dietary fibers (β-glucans). BSG contains around 17% of cellulose and 28% of non-cellulose polysaccharides (arabinoxylans and lignine). According to some research, BSG retains significant amounts of polyphenols [80–82]. Ktenioudaki et al. [83] published an article where brewer's spent grain can be used as functional food in baked snacks, specifically in breadsticks in this case. However, as reported by the authors, some baking properties were significantly affected and this can be correlated with the high fiber content in BSG. Namely, breadsticks containing BSG were darker, less crispy, and of a lower baking volume. Nevertheless, the quality of breadsticks after 50 days of storage remained unchanged.

From an environmental point of view, the elimination of industrial by-products represents a sane solution to pollution problems and deserves attention.

Much research has been conducted and is still ongoing to reveal and determine the chemistry and biology of mycotoxin production in fungi on many cereals. However, the literature data on mycotoxins in wheat malt and beer are modest and even less data are found on mycotoxins in malting and brewing by-products. Therefore, it is relevant to research and monitor the occurrence of these mycotoxins in wheat food and feed, including malting and brewing products and by-products.

While the FDA proposed a rule on Good Manufacturing Practice, which suggested that separating and selling by-products derived from malting and brewing industries should obligate the manufacturers to honor food safety regulations, the Beer Institute and American Malting Barley Association Inc. strongly disagreed because this would mean additional costs for maltsters and brewers on which they could not agree upon [84].

7. The Disposal/Use of Infected Grains

Considering all the risks infected grains can do to the economy in malting and brewing industries, the end question is what to do with infected and contaminated grains. This is rarely mentioned in the scientific papers, and is of great concern for growers, farmers, and M/B industries. However, there are some operations that can reduce the damage caused by fungal infection. Namely, diluting the contaminated grains (only for feed) with healthy ones, removing the outer layer of contaminated grain, and applying chemical and physical decontamination can lessen the detrimental consequences of fungal contamination. Alkaline treatments, binders that can reduce the bioavailability of mycotoxins in the intestinal tract, detoxifiers, and enzymes can be beneficial to toxins' reduction. Biocontrol is also a possible method. However, the question remains: What to do with heavily infected *Fusarium* grains? Handling the infected grains demands appropriate equipment since grain dust can contain DON and other toxins or fungal spores. During storage and handling, such grains should be excluded from healthy ones and one should be careful not to disperse the spores in the local fields [85]. Badly infected grains could be burnt or possibly used in digestion for biogas or bioethanol production. Composting is also an option. Agnew and Kirsch [86] reported the results from their project, where they discovered that composting seemed to be the safest and most efficient way to eliminate DON from the grains. However, it should be taken into an account that other microorganisms from the composting pile probably modified DON into a less toxic form. Namely, modified forms appear to be less toxic, but when degraded in the gut, they reform into their basic toxic form.

In a nutshell, so far, the most acceptable methods for mycotoxins' decontamination are [87]:

a. Clean and blend—with the uncontaminated grain; exclusively for animal feed;
b. Dumping—grains with high toxin levels can be disposed in a bush, slough, or in a hole;
c. Gravity sorter—grains infected with *Fusarium* have a lower weight and are easy to sort with this technique;
d. Burning—the most common method of disposal. This requires an investment in a grain-burning stove, but the produced heat can be used for heating. However, according to Agnew and Kirsch [86], ash can still contain DON and fungal spores and thus represents a danger to human health;
e. Composting;
f. Anaerobic digestion; and
g. Gamma irradiation [88].

8. Conclusions

Literature data indicate that malting and brewing by-products withhold significant amounts of various mycotoxins and for that reason, attention should be paid to ensure the health of animal feed. It is necessary to monitor multi-(myco)toxin concentrations and also to expand the analysis spectrum on masked toxins, since masked mycotoxins can be degraded during digestion by lactic acid bacteria in the intestinal track and recover the toxicity of unmasked mycotoxin [47,89,90]. New masked forms of DON (DON sulphates) that can occur in grains naturally have been confirmed recently [89], but no occurrence or metabolism studies were made. However, in the several research papers Mastanjević et al. published [1,2], the problem of mycotoxins in malting and brewing by-products was referred to.

Mycotoxins are retained in germ/rootlets, spent grains, and spent yeast depending on the genotype susceptibility, starting mycotoxin concentrations, and fungicide treatment [91]. Since the spectrum of found mycotoxins is so broad, as confirmed by Mastanjević et al. [1], this calls for attention and updates on legislation to ensure the health of animals and, subsequently, humans.

According to the literature [68,69], suitable storage conditions for these by-products are indisputably important and should be provided (low a_w and temperature and shorter storage time).

Author Contributions: Writing review & Editing, K.M. (Kristina Mastanjević); Conceptualization, J.L.; Investigation, M.J. and B.Š.; Visualization, V.K.; Data Acquisition, K.M. (Kristina Mastanjević); Supervising, K.M. (Krešimir Mastanjević), all contributed equally.

Funding: This research received no external funding.

Acknowledgments: Many thanks to all the authors who took part in writing this manuscript and showed a professional and scientific approach to the subject.

Conflicts of Interest: The authors declare no conflict of interest.

References

1. Mastanjević, K.; Šarkanj, B.; Warth, B.; Krska, R.; Sulyok, M.; Mastanjević, K.; Šantek, B.; Krstanović, V. *Fusarium culmorum* multi-toxin screening in malting and brewing byproducts. *LWT* **2018**, *98*, 642–645. [CrossRef]
2. Mastanjević, K.; Šarkanj, B.; Šantek, B.; Mastanjević, K.; Krstanović, V. *Fusarium culmorum* mycotoxin transfer from wheat to malting and brewing products and by-products. *World Mycotoxin J.* **2018**, in press. [CrossRef]
3. MAGB. Available online: http://www.ukmalt.com/malting-co-products (accessed on 16 March 2018).
4. Mussato, S.I. Biotechnological potential of brewing industry by-products. In *Biotechnology for Agro-Industrial Residues Utilization*; Singh nee' Nigam, P., Pandey, A., Eds.; Springer: Dordrecht, The Netherlands, 2009; pp. 313–326.
5. Krstanović, V.; Klapec, T.; Velić, N.; Milaković, Z. Contamination of malt barley and wheat by *Fusarium graminearum* and *Fusarium culmorum* from the crop years 2001–2003 in Eastern Croatia. *Microbiol. Res.* **2005**, *160*, 353–359. [CrossRef]
6. Snijders, C.H.A. Systemic fungal growth of *Fusarium culmorum* in stems of winter wheat. *J. Phytopathol.* **1990**, *129*, 133–140. [CrossRef]
7. Parikka, P.; Hakala, K.; Tiilikkala, K. Expected shifts in *Fusarium* species' composition on cereal grain in Northern Europe due to climatic change. *Food Addit. Contam. A* **2012**, *29*, 1543–1555. [CrossRef]
8. Ward, T.J.; Clear, R.M.; Rooney, A.P.; O'Donnell, K.; Gaba, D.; Patrick, S.; Starkey, D.E.; Gilbert, J.; Geiser, D.M.; Nowicki, T.W. An adaptive evolutionary shift in Fusarium head blight pathogen populations is driving the rapid spread of more toxigenic *Fusarium graminearum* in North America. *Fungal Genet. Biol.* **2008**, *45*, 473–484. [CrossRef]
9. Zhang, H.; Van der Lee, T.; Waalwijk, C.; Chen, W.; Xu, J.; Xu, J.; Zhang, Y.; Feng, L. Population analysis of the *Fusarium graminearum* species complex from wheat in china show a shift to more aggressive isolates. *PLoS ONE* **2012**, *7*, e31722. [CrossRef]
10. Roháčik, T.; Hudec, K. Fungal infection of malt barley kernels in Slovak Republic. *Plant Prot.Sci.* **2007**, *43*, 86–93. [CrossRef]
11. Gonzalez Pereyra, M.L.; Rosa, C.A.R.; Dalcero, A.M.; Cavaglieri, L.R. Mycobiota and mycotoxins in malted barley and brewer's spent grain from Argentinean breweries. *Lett. Appl. Microbiol.* **2011**, *53*, 649–655. [CrossRef]
12. Piacentini, K.C.; Savi, G.D.; Pereira, M.E.V.; Scussel, V.M. Fungi and the natural occurrence of deoxynivalenol and fumonisins in malting barley (*Hordeum vulgare* L.). *Food Chem.* **2015**, *187*, 204–209. [CrossRef]
13. Mastanjević, K.; Krstanović, V.; Mastanjević, K.; Šarkanj, B. Malting and Brewing Industries Encounter *Fusarium* spp. Related Problems. *Fermentation* **2018**, *4*, 3. [CrossRef]
14. Habler, K.; Hofer, K.; Geissinger, C.; Schuler, J.; Huckelhoven, R.; Hess, M.; Gastl, M.; Rychlik, M. Fate of *Fusarium* toxins during the malting process. *J. Agric. Food Chem.* **2016**, *64*, 1377–1384. [CrossRef]
15. CAST (Councile for Agricultural Science and Technology). *Mycotoxins: Risk in Plant, Animal and Human Systems*; Task Force Report No. 139; CAST: Ames, Iowa, 2003.
16. Jarvis, B.B. Chemistry and toxicology of molds isolated from water-damaged buildings, Mycotoxins and Food Safety. *Adv. Exp. Med. Biol.* **2002**, *504*, 43–52.
17. Pennsylvania Mycotoxin Management Guidance Document. Deoxynivalenol (DON) in Corn. Pennsylvania Department of Agriculture, Pennsylvania Department of Environmental Protection, 2012. Available online: https://www.agriculture.pa.gov/Food/pet_food/Documents/ManagementofDONincorn.pdf (accessed on 8 November 2018).
18. Varga, E.; Wiesenberger, G.; Hametner, C.; Ward, T.J.; Dong, Y.; Schöfbeck, D.; McCormick, S.; Broz, K.; Stückler, R.; Schuhmacher, R.; et al. New tricks of an old enemy: Isolates of *Fusarium graminearum* produce a type A trichothecene mycotoxin. *Environ. Microbiol.* **2015**, *17*, 2588–2600. [CrossRef]
19. Berthiller, F.; Sulyok, M.; Krska, R.; Schuhmacher, R. Chromatographic methods for the simultaneous determination of mycotoxins and their conjugates in cereals. *Int. J. Food Microbiol.* **2007**, *119*, 33–37. [CrossRef]
20. Dupire, S. Mycotoxins and other contaminants in the malting and brewing industries. In Proceedings of the European Brewery Convention Congress, Dublin, Ireland, 17–22 May 2003.

21. Melotte, L. Survey on the analysis of mycotoxins. Analysis Committee of the European Brewery Convention, European Brewery Association. *J. Inst. Brew.* **2004**, *110*, 235–239.
22. Directive 2002/32/EC. OJEC, L140/10. Available online: https://eur-lex.europa.eu/resource.html?uri=cellar:aca28b8c-bf9d-444f-b470-268f71df28fb.0004.02/DOC_1&format=PDF (accessed on 8 January 2019).
23. Recommendation 2006/576/EC. OJEC, L229/7. Available online: https://eur-lex.europa.eu/LexUriServ/LexUriServ.do?uri=OJ:L:2006:229:0007:0009:EN:PDF (accessed on 8 January 2019).
24. Recommendation 2013/165/EC. OJEC L 91, 12–15. Available online: https://eur-lex.europa.eu/LexUriServ/LexUriServ.do?uri=OJ:L:2013:091:0012:0015:EN:PDF (accessed on 8 January 2019).
25. Ruiz-Medina, A.; Fernández-de Córdova, M.L. Aflatoxin B_1 in beer at different Stages of Production. In *Processing and Impact on Active Components in Food*; Preedy, V., Ed.; Elsevier/Academic Press: San Diego, CA, USA, 2015; pp. 517–523.
26. Commission Regulation 2174/2003 EC. OJEC, L326, 12–15. Available online: https://publications.europa.eu/en/publication-detail/-/publication/c849b56b-42ec-4e34-8c90-3a3668737420 (accessed on 8 January 2019).
27. World Health Organization International Agency for Research on Cancer IARC. Monographs on the Evaluation of Carcinogenic Risks to Humans Volume 82. 2002. Available online: http://monographs.iarc.fr/ENG/Monographs/vol82/mono82.pdf (accessed on 8 July 2018).
28. Mably, M.; Mankotia, M.; Cavlovic, P.; Tam, J.; Wong, L.; Pantazopoulos, P.; Calway, P.; Scott, P.M. Survey of aflatoxins in beer sold in Canada. *Food Addit. Contam.* **2005**, *22*, 1252–1257. [CrossRef]
29. Nakajima, M.; Tsubouchi, H.; Miyabe, M. A survey of ochratoxin A and aflatoxins in domestic and imported beers in Japan by immunoaffinity and liquid chromatography. *J. AOAC Int.* **1999**, *82*, 897–902.
30. Inoue, T.; Nagatomi, Y.; Uyama, A.; Mochizuki, N. Fate of mycotoxins during beer brewing and fermentation. *Biosci. Biotechnol. Biochem.* **2013**, *77*, 1410–1415. [CrossRef]
31. Reddy, K.R.N.; Salleh, B. A preliminary study on the occurrence of *Aspergillus* spp. and aflatoxin B_1 in imported wheat and barley in Penang, Malaysia. *Mycotoxin Res.* **2010**, *26*, 267–271. [CrossRef]
32. Park, J.W.; Kim, E.K.; Shon, D.H.; Kim, Y.B. Natural co-occurrence of aflatoxin B_1, fumonisin B_1 and ochratoxin A in barley and corn foods from Korea. *Food Addit. Contam.* **2002**, *19*, 1073–1080. [CrossRef]
33. Scudamore, K.A.; Patel, S. Survey for aflatoxins, ochratoxin A, zearalenone and fumonisins in maize imported into United Kingdom. *Food Addit. Contam.* **2000**, *17*, 407–416. [CrossRef]
34. Lewis, L.; Onsongo, M.; Njapau, H.; Shurz-Rogers, H.; Luber, G.; Kieszak, S.; Nyamongo, J.; Backer, L.; Dahiye, A.M.; Misore, A.; et al. Aflatoxin contamination of commercial maize products during an outbreak of acute aflatoxicosis in eastern and central Kenya. *Environ. Health Perspect.* **2005**, *113*, 1763–1767. [CrossRef]
35. Gerbaldo, G.A.; Pereyra, C.M.; Cavaglieri, L.R.; Ruiz, F.; Pascual, L.; Dalcero, A.M.; Barberis, A.M. Surveillance of aflatoxin and microbiota related to brewers grain destined for swine feed in Argentina. *Vet. Med. Int.* **2011**, 912480. [CrossRef]
36. Papadopoulou-Bouraoui, A.; Vrabcheva, T.; Valzacchi, S.; Stroka, J.; Anklam, E. Screening survey of deoxynivalenol in beer from the European market by an enzyme-linked immunosorbent assay. *Food Addit. Contam.* **2004**, *21*, 607–617. [CrossRef]
37. Milićević, D.R.; Škrinjar, M.; Baltić, T. Real and perceived risks for mycotoxin contamination in foods and feeds: Challenges for food safety control. *Toxins* **2010**, *2*, 527–592. [CrossRef]
38. Salas, B.; Steffenson, B.J.; Casper, H.H.; Tacke, B.; Prom, L.K.; Fetch, T.G., Jr.; Schwarz, P.B. *Fusarium* species pathogenic to barley and their associated mycotoxins. *Plant Dis.* **1999**, *83*, 667–674. [CrossRef]
39. Velić, N.; Pavlović, H.; Ćosić, J.; Kanižai, G.; Krstanović, V. A survey of *Fusarium graminearum* and deoxynivalenol contamination of malt barley from the crop year 2004 in eastern Croatia. *Cereal Res. Commun.* **2007**, *35*, 1293–1296. [CrossRef]
40. Legzdina, l.; Buerstmayr, H. Comparison of infection with Fusarium head blight and accumulation of mycotoxins in grain of hulles and covered barley. *J. Cereal Sci.* **2004**, *40*, 61–67. [CrossRef]
41. Rotter, B.A.; Prelusky, D.B. Toxicology of deoxynivalenol (vomitoxin). *J. Toxicol. Environ. Health* **1996**, *48*, 1–34. [CrossRef]
42. Saenz de Rodrigues, C.A.; Bongiovanni, A.M.; Conde de Borrego, L. An epidemic of precocious development in Puerto Rican children. *J. Pediatr.* **1985**, *107*, 393–396. [CrossRef]
43. Campbell, G.D. Trichothecene mycotoxicosis B a new entity? *S. Afr. Med. J.* **1991**, *80*, 361–362.

44. Dall'Erta, A.; Cirlini, M.; Dall'Asta, M.; Del Rio, D.; Galaverna, G.; Dall'Asta, C. Masked mycotoxins are efficiently hydrolyzed by human colonic microbiota releasing their aglycones. *Chem. Res. Toxicol.* **2013**, *26*, 305–312. [CrossRef]
45. Berthiller, F.; Werner, U.; Sulyok, M.; Krska, R.; Hauser, M.-T.; Schuhmacher, R. Liquid chromatography coupled to tandem mass spectrometry (LC-MS/MS) determination of phase II metabolites of the mycotoxin zearalenone in the model plant *Arabidopsis thaliana*. *Food Addit. Contam.* **2006**, *23*, 1194–1200. [CrossRef]
46. González Pereyra, M.L.; Sulyok, M.; Baralla, V.; Dalcero, A.M.; Krska, R.; Chulze, S.; Cavaglieri, L.R. Evaluation of zearalenone, -zearalenol, -zearalenol, zearalenone 4-sulfate and -zearalenol 4-glucoside levels during the ensiling process. *World Mycotoxin J.* **2014**, *7*, 291–295. [CrossRef]
47. Berthiller, F.; Crews, C.; Dall'Asta, C.; Saeger, S.D.; Haesaert, G.; Karlovsky, P.; Oswald, I.P.; Seefelder, W.; Speijers, G.; Stroka, J. Masked mycotoxins: A review. *Mol. Nutr. Food Res.* **2013**, *57*, 165–186. [CrossRef]
48. Binder, S.B.; Schwartz-Zimmermann, H.E.; Varga, E.; Bichl, G.; Michlmayr, H.; Adam, G.; Berthiller, F. Metabolism of zearalenone and its major modified forms in pigs. *Toxins* **2017**, *9*, 56. [CrossRef]
49. Dall'Asta, C.; Battilani, P. Fumonisins and their modified forms, a matter of concern in future scenario? *World Mycotoxin J.* **2016**, *9*, 727–739. [CrossRef]
50. Scott, P.M.; Lawrence, G.A. Analysis of beer for fumonisins. *J. Food Prot.* **1995**, *58*, 1379–1382. [CrossRef]
51. Daško, L'.; Rauová, D.; Belajová, E.; Kováč, M. Determination of fumonisins B_1 and B_2 in beer. *Czech. J. Food Sci.* **2005**, *23*, 20–26. [CrossRef]
52. Voss, K.A.; Riley, R.T. Fumonisin toxicity and mechanism of action: Overview and current perspectives. *Food Saf.* **2013**, *1*, 49–69. [CrossRef]
53. Scott, P.M. Recent research on fumonisins: A review. *Food Addit. Contam. A* **2012**, *29*, 242–248. [CrossRef]
54. Missmer, S.A.; Suarez, L.; Felkner, M.; Wang, E.; Merrill, A.H., Jr.; Rothman, K.J.; Hendricks, K.A. Exposure to fumonisins and the occurrence of neural tube defects along the Texas-Mexico border. *Environ. Health Perspect.* **2006**, *114*, 237–241. [CrossRef]
55. Rheeder, J.; Marasas, W.; Theil, P.; Sydenham, E.; Shephard, G.; Van Schalkwyk, D. *Fusarium moniliforme* and fumonisins in corn in relation to human esophageal cancer in Transkei. *Phytopathology* **1992**, *82*, 353–357. [CrossRef]
56. Braun, M.S.; Wink, M. Exposure, occurrence, and chemistry of fumonisins and their cryptic derivatives. *Compr. Rev. Food Sci. Food Saf.* **2018**, *17*, 769–791. [CrossRef]
57. Maranghi, F.; Tassinari, R.; Narciso, L.; Tait, S.; Rocca, C.L.; Felice, G.D.; Butteroni, C.; Corinti, S.; Barletta, B.; Cordelli, E.; et al. In Vivo Toxicity and Genotoxicity of Beauvericin and Enniatins. Combined Approach to Study In Vivo Toxicity and Genotoxicity of Mycotoxins Beauvericin (BEA) and Enniatin B (ENNB). *EFSA Supporting Publication* **2018**, *15*, 1406E. Available online: https://efsa.onlinelibrary.wiley.com/doi/epdf/10.2903/sp.efsa.2018.EN-1406 (accessed on 8 January 2019). [CrossRef]
58. Anli, E.; Mert Alkis, I. Ochratoxin A and brewing technology: A Review. *J. Inst. Brew.* **2010**, *116*, 23–32. [CrossRef]
59. Creppy, E.E. Human ochratoxicosis. *J. Toxicol.* **1999**, *18*, 277–293. [CrossRef]
60. Huff, W.E.; Doerr, J.A. Synergism between aflatoxin and ochratoxin A in broiler chickens. *Poult. Sci.* **1981**, *60*, 550–555. [CrossRef]
61. Peraica, M.; Rašić, D. The impact of mycotoxicoses on human history. *Arch. Ind. Hyg. Toxicol.* **2012**, *63*, 513–518. [CrossRef]
62. Veprikova, Z.; Vaclavikova, M.; Lacina, O.; Dzuman, Z.; Zachariasova, M.; Hajslova, J. Occurrence of mono and di-glycosylated conjugates of T-2 and HT-2 toxins in naturally contaminated cereals. *World Mycotoxin J.* **2012**, *5*, 231–240. [CrossRef]
63. Nakagawa, H.; Sakamoto, S.; Sago, Y.; Nagashima, H. Detection of type A trichothecene di-glucosides produced in corn by high-resolution liquid chromatography-orbitrap mass spectrometry. *Toxins* **2013**, *5*, 590–604. [CrossRef]
64. McCormick, S.P.; Price, N.P.J.; Kurtzman, C.P. Glucosylation and other biotransformations of T-2 toxin by yeasts of the *Trichomonascus* clade. *Appl. Environ. Microbiol.* **2012**, *78*, 8694–8702. [CrossRef]
65. Nathanail, A.V.; Varga, E.; Meng-Reiterer, J.; Bueschl, C.; Michlmayr, H.; Malachova, A.; Fruhmann, P.; Jestoi, M.; Peltonen, K.; Adam, G.; et al. Metabolism of the *Fusarium* mycotoxins T-2 toxin and HT-2 toxin in wheat. *J. Agric. Food Chem.* **2015**, *63*, 7862–7872. [CrossRef]

66. Wen-Shyg Chiou, P.; Chen, C.R.; Chen, K.J.; Yu, B. Wet brewers' grains or bean curd pomance as partial replacement of soybean meal for lactating cows. *Anim. Feed Sci. Technol.* **1998**, *74*, 123–134. [CrossRef]
67. Cavaglieri, L.R.; Keller, K.M.; Pereya, C.M.; Gonzalez Pereya, M.L.; Alonso, V.A.; Rojo, F.G.; Dalcero, A.M.; Rosa, C.A.R. Fungi and natural incidence of selected mycotoxins in barley rootlets. *J. Stored Prod. Res.* **2009**, *45*, 147–150. [CrossRef]
68. Kensler, T.W.; Roebuck, B.D.; Wogan, G.N.; Groopman, J.D. Aflatoxin: A 50-year odyssey of mechanistic and translational toxicology. *Toxicol. Sci.* **2011**, *120*, S28–S48. [CrossRef]
69. Habschied, K.; Šarkanj, B.; Klapec, T.; Krstanović, V. Distribution of zearalenone in malted barley fractions dependent on *Fusarium graminearum* growing conditions. *Food Chem.* **2011**, *129*, 329–332. [CrossRef]
70. Habschied, K.; Šarkanj, B.; Krstanović, V.; Velić, N.; Novak, M.; Mastanjević, K. *Fusarium* mycotoxins in malting and brewing by-products. In Proceedings of the World Mycotoxin Forum 8th Conference, Mycotoxin Control: The System Approach, Vienna, Austria, 10–12 November 2014; p. 85.
71. Krstanović, V.; Mastanjević, K.; Velić, N.; Pleadin, J.; Perši, N.; Španić, V. The influence of *Fusarium culmorum* contamination level on deoxynivalenol content in wheat, malt and beer. *Roman. Biotechnol. Lett.* **2015**, *20*, 10901–10910.
72. Blezinger, S.B. Feed Supplements Come in Several Different Forms: Part 4. Cattle Today Online. Available online: http://www.cattletoday.com/archive/2003/February/CT251.shtml (accessed on 16 March 2018).
73. Ha, J.-K.; Kim, S.W.; Kim, W.Y. *Use of Agro-Industrial by-Products as Animal Feeds in Korea*; ASPAC Food & Fertilizer Technology Center: Taipei, Taiwan, 1996; pp. 1–15.
74. Rychlik, M.; Humpf, H.U.; Marko, D.; Dänicke, S.; Mally, A.; Berthiller, F.; Klaffke, H.; Lorenz, N. Proposal of a comprehensive definition of modified and other forms of mycotoxins including "masked" mycotoxins. *Mycotoxin Res.* **2014**, *30*, 197–205. [CrossRef]
75. Maul, R.; Pielhau, R.; Koch, M. Evaluation of an extraction method and spin column cleanup procedure for *Fusarium* mycotoxins and their masked derivatives from grain matrix. *Food Control* **2014**, *40*, 151–156. [CrossRef]
76. De Boevre, M.; Di Mavungu, J.D.; Landschoot, S.; Audenaert, K.; Eeckhout, M.; Maene, P.; Haesaert, G.; De Saeger, S. Natural occurrence of mycotoxins and their masked forms in food and feed products. *World Mycotoxin J.* **2012**, *5*, 207–219. [CrossRef]
77. Zachariasova, M.; Dzuman, Z.; Veprikova, Z.; Hajkova, K.; Jiru, M.; Vaclavikova, M.; Pospichalova, M.; Florian, M.; Hajslova, J. Occurrence of multiple mycotoxins in European feedingstuffs, assessment of dietary intake by farm animals. *Anim. Feed Sci. Technol.* **2014**, *193*, 124–140. [CrossRef]
78. Jozinović, A.; Šubarić, D.; Ačkar, Đ.; Miličević, B.; Babić, J.; Jašić, M.; Lendić, V.K. Food industry by-products as raw materials in functional food production. *Hrana u Zdravlju i Bolesti* **2014**, *3*, 22–30.
79. Obradović, V.; Babić, J.; Šubarić, D.; Ačkar, Đ.; Jozinović, A. Improvement of nutritional and functional properties of extruded food products. *J. Food Nutr. Res.* **2014**, *53*, 189–206.
80. Meneses, N.G.T.; Martins, S.; Teixeira, J.A.; Mussatto, S.I. Influence of extraction solvents on the recovery of antioxidant phenolic compounds from brewer's spent grains. *Sep. Purif. Technol.* **2013**, *108*, 152–158. [CrossRef]
81. Moreira, M.M.; Morais, S.; Carvalho, D.O.; Barros, A.A.; Delerue-Matos, C.; Guido, L.F. Brewer's spent grain from different types of malt: Evaluation of the antioxidant activity and identification of the major phenolic compounds. *Food Res. Int.* **2013**, *54*, 382–388. [CrossRef]
82. McCarthy, A.L.; O'Callaghan, C.; Connolly, A.; Piggott, C.O.; FitzGerald, R.J.; O'Brien, N.M. Phenolic extracts of brewers' spent grain (BSG) as functional ingredients-Assessment of their DNA protective effect against oxidant-induced DNA single strand breaks in U937 cells. *Food Chem.* **2012**, *134*, 641–646. [CrossRef]
83. Ktenioudaki, A.; Chaurin, V.; Reis, S.; Gallagher, E. Brewer's spent grain as a functional ingredient for breadsticks. *Int. J. Food Sci. Technol.* **2012**, *47*, 1765–1771. [CrossRef]
84. Corwell, C. Spent Grain Regulation: Commentary from the Beer Institute. 2014. Available online: https://www.craftbrewingbusiness.com/business-marketing/spent-grain-regulation-commentary-beer-institute/ (accessed on 16 March 2018).
85. Scott, T. Feed and Processing Options for Heavily Downgraded Wheat. 2014. Available online: https://cigi.ca/wp-content/uploads/2014/11/Tom-Scott-Feed-and-Processing-Options-for-Heavily-Downgraded-Wheat.pdf (accessed on 12 December 2017).

86. Agnew, J.; Kirsch, J. Evaluation of Disposal Options for *Fusarium* Damaged Grain and Screenings. 2017. Available online: http://pami.ca/wp-content/uploads/2017/04/R8415-Research-Report.pdf (accessed on 16 March 2018).
87. King, C. Extracting Value from Fusariem-Damage Grain and Screenings. 2017. Available online: https://www.topcropmanager.com/diseases/from-waste-to-worth-20793 (accessed on 4 August 2018).
88. Snizhko, O.; Lomova, N.; Narizhnyy, S.; Mingaleeva, Z. Enhancing food safety of pollen by means of irradiation. *Ukr. Food J.* **2015**, *4*, 32–39. Available online: http://ufj.ho.ua/Archiv/UKRAINIAN%20FOOD%20JOURNAL%202015%20V.4%20Is.1.pdf (accessed on 15 November 2018).
89. Berthiller, F.; Krska, R.; Domig, K.J.; Kneifel, W.; Juge, N.; Schuhmacher, R.; Adam, G. Hydrolytic fate of deoxynivalenol-3-glycoside during digestion. *Toxicol. Lett.* **2011**, *206*, 264–267. [CrossRef] [PubMed]
90. Warth, B.; Fruhmann, P.; Wiesenberger, G.; Kluger, B.; Sarkanj, B.; Lemmens, M.; Hametner, C.; Fröhlich, J.; Adam, G.; Krska, R.; et al. Deoxynivalenol-sulfates: Identification and quantification of novel conjugated (masked) mycotoxins in wheat. *Anal. Bioanal. Chem.* **2015**, *407*, 1033–1039. [CrossRef]
91. Mastanjević, K.; Španić, V.; Horvat, D.; Mastanjević, K.; Šarkanj, B.; Krstanović, V.; Šantek, B. Establishing the impact of *Fusarium culmorum* infection and fungicide treatment on wheat malt quality. *J. Food Process. Preserv.* **2018**, *42*, e13714. [CrossRef]

toxins

MDPI

Article

Detection of Shiga Toxin 2 Produced by *Escherichia coli* in Foods Using a Novel AlphaLISA

Cheryl M. Armstrong [1], Leah E. Ruth [2], Joseph A. Capobianco [1], Terence P. Strobaugh Jr. [1], Fernando M. Rubio [2] and Andrew G. Gehring [1,*]

1 Molecular Characterization of Foodborne Pathogens Research Unit, United States Department of Agriculture, Eastern Regional Research Center, Wyndmoor, PA 19038, USA; Cheryl.Armstrong@ars.usda.gov (C.M.A.); Joseph.Capobianco@ars.usda.gov (J.A.C.); Terence.Strobaugh@ars.usda.gov (T.P.S.J.)

2 Abraxis, Inc., Warminster, PA 18974, USA; lruth@abraxiskits.com (L.E.R.); frubio@abraxiskits.com (F.M.R.)

* Correspondence: Andrew.Gehring@ars.usda.gov; Tel.: +1-215-233-6491

Received: 14 September 2018; Accepted: 17 October 2018; Published: 23 October 2018

Abstract: Amplified luminescent proximity homogenous assay-linked immunosorbent assay (AlphaLISA) is comprised of a bead-based immunoassay that is used for small molecule detection. In this study, a novel AlphaLISA was developed and optimized for the detection of Shiga-toxin 2 (Stx2). Efficacy and sensitivity trials showed the AlphaLISA could detect ≥0.5 ng/mL of purified Stx2, which was comparable to the industry-standard enzyme-linked immunosorbent assay (ELISA) tests for Stx2 detection. In addition, evaluation of Shiga toxin-producing *Escherichia coli* (STEC)-inoculated Romaine lettuce and ground beef samples demonstrated that both the AlphaLISA and the ELISA were able to discern uninoculated samples from 1× and 10× diluted samples containing ~10 CFU/mL of STEC enriched in modified tryptic soy broth with mitomycin C for 16 h. Overall, the increased signal-to-noise ratios indicated a more robust signal was produced by the AlphaLISA compared to the ELISA and the delineation of higher toxin concentrations without the need for sample dilution implied a greater dynamic range for the AlphaLISA. Implementation of the newly developed AlphaLISA will allow for more rapid analysis for Stx2 with less manual manipulation, thus improving assay throughput and the ability to automate sample screening while maintaining detection limits of 0.5 ng/mL.

Keywords: amplified luminescent proximity homogenous assay-linked immunosorbent assay (AlphaLISA); detection; enzyme-linked immunosorbent assay (ELISA); *E. coli*; Shiga toxin; STEC; Stx2

Key Contribution: There are many benefits to the newly developed AlphaLISA including a shorter and simpler homogeneous protocol that requires less manual manipulation and no wash steps compared to the ELISA. Overall, this AlphaLISA allows for the rapid detection of bacterial-generated Stx2 in food matrices using a highly sensitive, robust, and automatable process that is amendable to high-throughput screening.

---

1. Introduction

Food poisoning due to Shiga toxin-producing *Escherichia coli* (STEC) is a consistent cause for concern in the United States because of its association with hemorrhagic colitis. Clinically, hemorrhagic colitis is characterized by the onset of a variety of symptoms including nausea, vomiting, abdominal pain, diarrhea, and bloody stools with approximately 5% of cases progressing to a more severe form of clinical disease known as hemolytic uremic syndrome (HUS) [1]. Expenses related to STEC illnesses are difficult to quantify; however, estimates from summary health measures such as the annual number of illnesses, hospitalizations, and deaths as estimated by the Centers for Disease Control and Prevention

(CDC) [2] can be used to approximate the costs to be around $300 M and a loss of 1700 quality-adjusted life years annually [3].

Rapid and accurate identification of STEC is imperative for the protection of several facets of human health. For example, stopping STEC contamination in food production is critical for curtailing the number of infections and preventing full-scale outbreaks from occurring, while obtaining an accurate diagnosis upon infection is crucial for ensuring proper care for patients. Because treatment of STEC infections with antibiotics greatly increases the risk of serious complications resulting from the development of the condition known as hemolytic uremic syndrome (HUS) [4,5] and Shiga toxin is known to be essential for the development of HUS from STEC infections [6]; the production and subsequent release of Shiga toxin has been a topic of intense investigation. Production of Shiga toxin is both a distinguishing feature and an important virulence factor for STEC strains. It can be expressed by strains carrying the *stx* gene, which is encoded for by a lambdoid bacteriophage [7,8]. Current research separates Shiga toxin into two main antigenically distinct groups, Shiga toxin 1 (Stx1) and Shiga toxin 2 (Stx2) [9]. Both toxins consist of a single A-subunit of ~32 kDa and five B-subunits of ~7.7 kDa each [10]. However, marked differences in the toxicity level exist between the two groups. For example, Stx2 types have an LD50 approximately 400 times lower than Stx1 [11] despite having ~60% identity at the amino acid level between the two groups [10]. Each of the two main groups can be further subdivided into the following variants: Stx1a, Stx1c, Stx1d, and Stx2a, Stx2b, Stx2c, Stx2d, Stx2e, Stx2f, and Stx2g [12]. The Food Safety and Inspection Service, an agency of the United States Department of Agriculture, has undertaken routine screening of meat samples for the presence of *stx* in order to identify the presence/absence of STEC in the food supply chain [13,14].

Because culture-based STEC detection can be long and laborious, alternative methods for identifying strains that produce Shiga toxin have emerged. More rapid antibody-based detection tests, specifically enzyme-linked immunosorbent assays (ELISAs) [15,16], antibody-based lateral flow assays (LATs) [17], and immunomagnetic separation assays (IMS) [18,19] have been employed as common testing platforms for the detection of STECs and Shiga toxin in lieu of traditional microbiological detection methods such as colony plating on selective media. A new platform known as the AlphaLISA, or amplified luminescent proximity homogenous assay-linked immunosorbent assay, utilizes both bead- and antibody-based technologies for pathogen or toxin detection. In this system, two different antibody-coated beads (aka. "donor" and "acceptor" beads) are used to bind either the same antigenic region on a target or two distinct antigenic regions of a target that are within the vicinity of one another. The donor beads contain a photosensitizing agent, which upon laser excitation at 680 nm, causes the donor bead to emit ~60,000 singlet oxygen molecules per second [20]. These singlet oxygen molecules then react with thioxene derivatives within the acceptor beads, resulting in the production of a chemiluminescent signal at 340–350 nm [21]. This signal activates fluorophores, also contained within the acceptor beads, resulting in a detectable signal consisting of emitted fluorescence with a narrow bandwidth centered around 615 nm. Because the half-life of the singlet oxygen is 4 microseconds, its diffusion distance is limited to approximately 200 nm in aqueous solutions. Thus, if the beads are not in close enough proximity, the singlet oxygen molecules from the donor bead decay to their ground state prior to reaching the acceptor beads and no signal is emitted.

The AlphaLISA has many advantages over the ELISA; namely its high sensitivity, relatively quick testing time, reduced hands-on workflow resulting from the ability to sequentially overlay the reagents, and it is easily adaptable to automation and high-throughput screening [22]. In this study, the first AlphaLISA for the detection of a bacterial toxin in foods was developed using the presence of Shiga toxin 2 (Stx2) as a marker for STEC contamination in foods. Comparisons were made between the newly developed AlphaLISA and an industry-standard ELISA in two different food matrices, Romaine lettuce and ground beef, to further evaluate the assay's utility as a rapid detection method.

2. Results

2.1. Antibody Configuration, Titration, and Gain Settings for Amplified Luminescent Proximity Homogenous Assay-Linked Immunosorbant Assay (AlphaLISA)

The AlphaLISA design presented here (Figure 1) utilizes two different monoclonal antibodies with specific affinity to Stx2 for incorporation onto the donor and acceptor beads. Antibody M-1005 binds the Shiga toxin 2 A-subunit while M-1003 binds the Shiga toxin 2 B-subunits. These antibodies were chosen so that the two beads would bind to different areas of the toxin, as antibodies with the same target site may compete with each other rather than trap the toxin between them. During assay development, antibody M-1003 was first used as the biotinylated antibody for association with the streptavidin-coated donor bead while the M-1005 antibody was conjugated to the acceptor bead (Figure 1A). Then, antibody M-1003 was conjugated to the acceptor bead while the M-1005 antibody was used as the biotinylated antibody for association with the streptavidin-coated donor bead (Figure 1B). In addition to testing both antibody configurations, the final concentration of the biotinylated antibody was also varied (0, 0.3, 1.0, and 3.0 nM) while keeping the acceptor-conjugated antibody beads constant to ensure optimal performance would be achieved for the detection of Stx2 in the AlphaLISA format. Results of this assay (Figure 2) showed that using 0.3 nM of biotinylated M-1005 antibody for association with the donor, while conjugating antibody M-1003 to the acceptor produced the largest signal intensities. Furthermore, this condition demonstrated the ability to differentiate between all levels of Stx2 as the *p*-values of independent Student's *t*-tests were less than 0.003. Since the response for this condition displayed the largest signal amplitudes and the ability to differentiate between each level of Stx2 tested, this configuration was subsequently used in the remainder of the assays presented. The final parameter examined in an effort to maximize the sensitivity and dynamic range of the AlphaLISA was the gain setting for the photomultiplier tube on the plate reader. A multifactorial test was performed using 0.3 nM of biotinylated M-1005 as the donor and conjugating M-1003 to the acceptor bead in an assay evaluating four levels of Stx2 (0, 3, 30, and 300 ng/mL) at three different gain settings (gain = 100, 125, and 150). While we observed larger signal intensities with the higher gains, we also noted that for some conditions, the assay was unable to differentiate between levels of Stx2 using a Student's *t*-test with an alpha value of 0.05. From this, it was determined that a gain setting of 100 was optimal because it demonstrated the greatest precision and was the best at resolving the lower levels of toxin (data not shown).

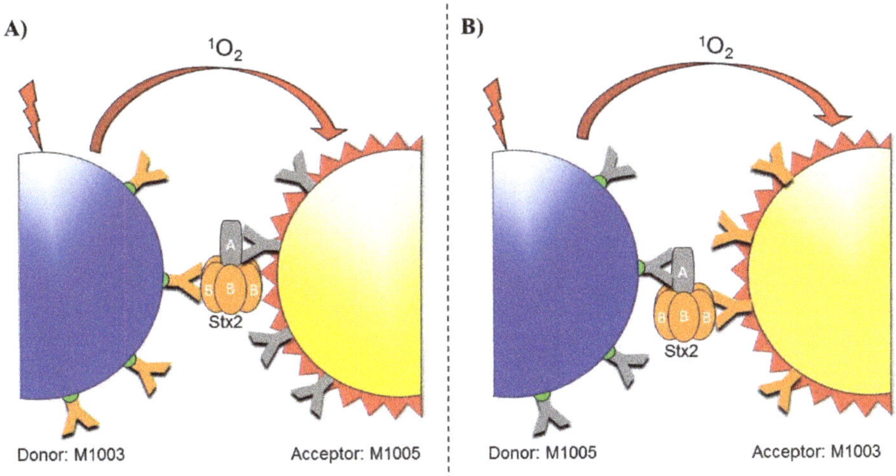

Figure 1. Schematic representation of the sandwich amplified luminescent proximity homogenous

assay-linked immunosorbent assay (AlphaLISA) for Shiga toxin detection. Two sets of monoclonal antibodies were employed in this assay (orange and gray Y's): one specific for the (**A**) subunit (M-1005) and other for the (**B**) subunit (M-1003) of Stx2. One set of antibodies was biotinylated (green hemisphere) to allow association with streptavidin-coated donor beads (blue sphere), while the other antibody set was directly conjugated to the acceptor beads (yellow sphere). When Stx2 is present, the donor and acceptor beads are colocalized through binding of the antibodies to the different subunits of the common antigen. Upon excitation by an Alpha laser at 680 nm (lightning bolt), the donor beads emit singlet oxygen molecules that react with the acceptor beads. Ultimately, energy transfer results in the emission of light by the acceptor beads, thus creating a detectable fluorescence signal at 615 nm. The following two formats were analyzed during assay development: (**A**) Antibody M-1003 was biotinylated and thus associated with the donor beads while M-1005 was conjugated to the acceptor beads and (**B**) antibody M-1005 was biotinylated and thus associated with the donor beads while M-1003 was conjugated to the acceptor beads.

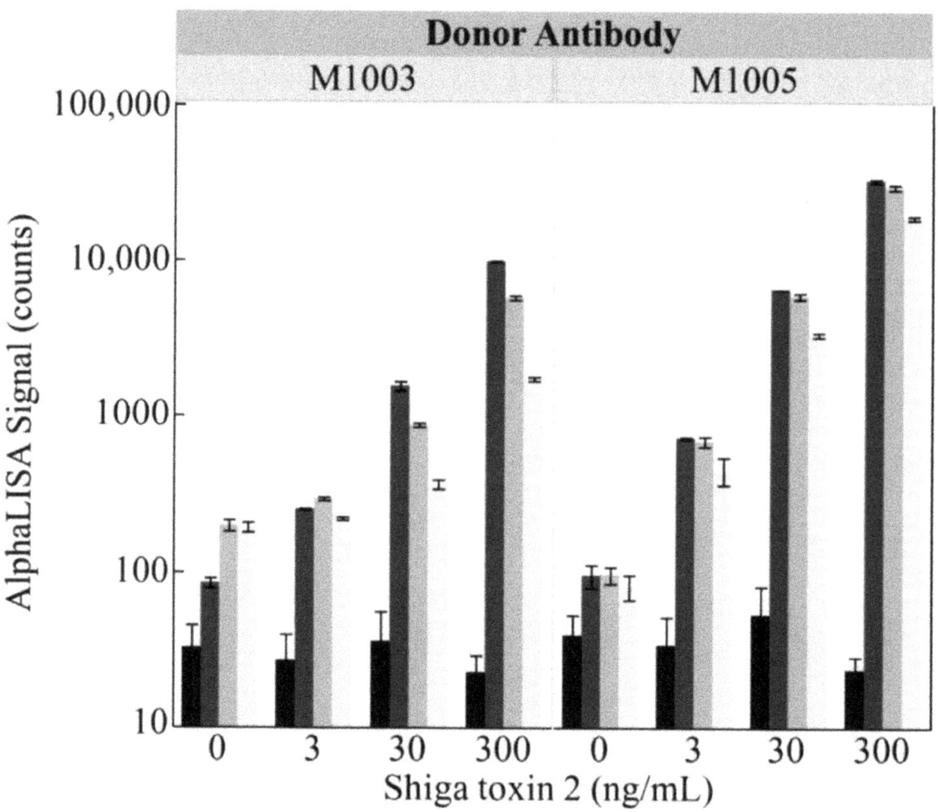

Figure 2. Optimization of donor/acceptor bead pairing and antibody concentration. A multifactorial test involving the donor/acceptor bead pairing, donor antibody concentration, and Stx2 concentration was performed to identify the optimal assay parameters for the AlphaLISA. AlphaLISA signals were recorded for the following factors: (a) bead pairings consisting of donor M-1003 with 25 μg acceptor M-1005 (left panel) versus donor M-1005 with 25 μg acceptor M-1003 (right panel); (b) donor antibody concentrations of 0 (black bars), 0.3 (dark gray bars), 1.0 (medium gray bars), and 3.0 nM (light gray bars); and (c) Stx2 concentrations of 0, 3 ng/mL, 30 ng/mL, and 300 ng/mL. Emitted light was measured on a BioTek Cytation 5 in Alpha mode at a gain setting of 100. Bars represent the mean values (n = 4) with error bars denoting the standard deviation.

2.2. *Sensitivity of the AlphaLISA Compared to the Enzyme-Linked Immunosorbent Assay (ELISA)*

Using the optimized parameters, purified Stx2 was then used to determine the sensitivity for the newly developed AlphaLISA. Varying concentrations of Stx2 (0, 0.5, 1, 3, and 10 ng/mL) were assayed using the AlphaLISA and the results recorded (Figure 3 black line). To benchmark the newly developed AlphaLISA against similar technologies, the same concentrations of Stx2 were also assayed using a commercially available ELISA (Figure 3 gray line). Six independent trials for each assay were conducted and the results were analyzed via independent Student's *t*-tests with alpha <0.05. The analysis demonstrated the ability of both the AlphaLISA and the ELISA to differentiate samples that do not contain Stx2 from samples containing 0.5, 1, 3, and 10 ng/mL of Stx2. In addition, independent *t*-tests (alpha < 0.05) demonstrated the ability of the assay to differentiate between all levels of Stx2 concentrations. Although the signal for both the AlphaLISA and the ELISA increased as the concentration of Stx2 increased, it is worth noting that the signal for the ELISA appeared to reach the maximum absorbance at 3 ng/mL of Stx2. (Note the maximum absorbance signal measurable by the instrument is 4.0.) The AlphaLISA, however, continued to show a marked increase in signal from 3 to 10 ng/mL.

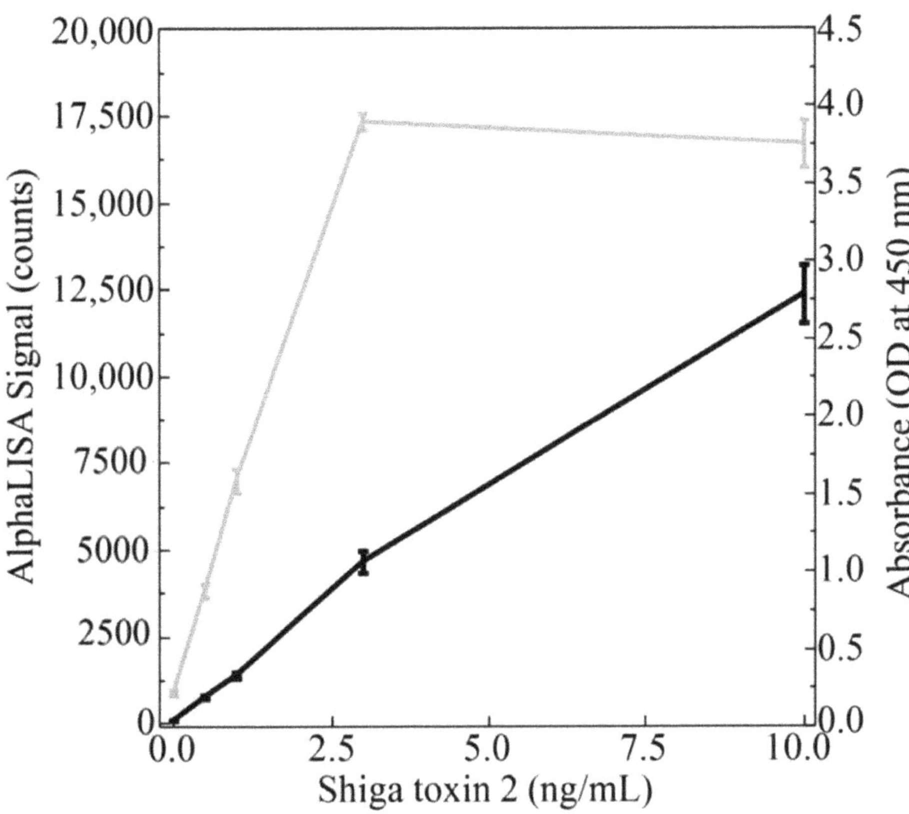

Figure 3. Detection of purified Stx2 using the AlphaLISA and the enzyme-linked immunosorbent assay (ELISA). Varying concentrations of purified Stx2 ranging from 0–10 ng/mL were analyzed using both the AlphaLISA (black line) and the ELISA (gray line). The response from both assays were quantified on a BioTek Cytation 5 with the emitted light from the AlphaLISA measured via the Alpha mode at a gain setting of 100 (left *y*-axis) and the absorbance values of the ELISA measured at 450 nm (right *y*-axis). Mean values from 6 trials were plotted with error bars denoting the standard deviation.

2.3. Detection of Stx2 in Food Matrices Using the AlphaLISA

In addition to the testing performed on purified toxin, the ability of the AlphaLISA to detect the presence of Stx2 in inoculated lettuce and ground beef samples was also examined to ascertain the functionality of the AlphaLISA in food matrices. These two food matrices were chosen because of their previous implication in STEC outbreaks [23–26]. For comparison purposes, lettuce inoculated with 9 CFU/mL of a Shiga toxin 2-producing *E. coli* O145 strain was tested by both AlphaLISA and ELISA for the presence of Stx2. During an overnight incubation, expression of Stx2 was induced in samples through the addition of mitomycin C. Samples consisting of 1× (undiluted) material, a 10× dilution, or a 100× dilution were then assayed (Figure 4A). Lettuce samples that had not been inoculated with STEC but were incubated in the same broth and under the same conditions were used as controls for this study. Because the output of the ELISA and the AlphaLISA differ, with the ELISA being a colorimetric assay measuring absorbance while the AlphaLISA yields emitted fluorescence, the numeric values are not directly comparable. However, both assays demonstrate a dose-dependent response. Both the AlphaLISA and ELISA were repeated 3 times in duplicate, and the data was analyzed using Student's *t*-tests, alpha = 0.05. In both the AlphaLISA and the ELISA, the 1× lysate dilution, the 10× lysate dilution and the control, were significantly different ($p < 0.0001$). However, the 100× lysate dilution could not be differentiated from the uninoculated control in either the AlphaLISA or the ELISA as $p = 0.73$ and 0.34, respectively. This implies that the AlphaLISA and the ELISA displayed similar Stx2 detection capabilities in lettuce.

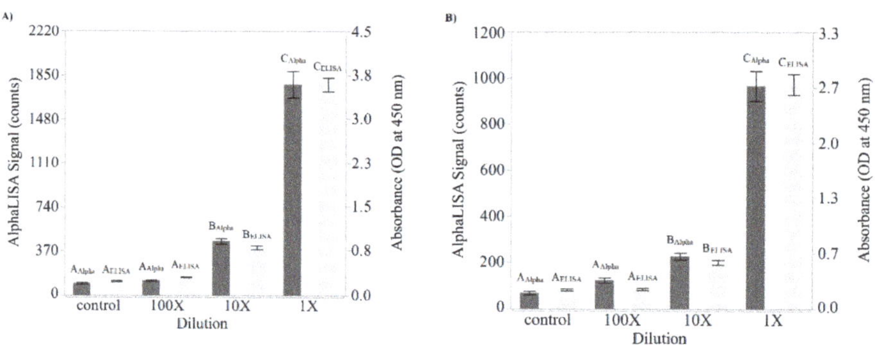

Figure 4. Detection of Stx2 in Shiga toxin-producing *Escherichia coli* (STEC)-inoculated foods using the AlphaLISA and ELISA. Stx2 production was induced using mitomycin C in STEC-inoculated (**A**) Romaine lettuce and (**B**) ground beef. Undiluted (1×) or diluted (10× and 100×) samples were subsequently assayed for the presence of Stx2 using both the AlphaLISA (dark gray bars) and the ELISA (light gray bars). Control samples contained uninoculated lettuce and ground beef samples, respectively. Responses were quantified on a BioTek Cytation 5 with the emitted light from the AlphaLISA measured via Alpha mode at a gain setting of 100 (left *y*-axis) and the absorbance values of the ELISA measured at 450 nm (right *y*-axis). Mean values from 3 independent trials containing 2 replicates were plotted with error bars denoting the standard deviation. Significance between AlphaLISA signal or absorbance values are denoted by dissimilar letters as determined by a Student's *t*-test at a 95% confidence level.

Testing of the inoculated ground beef samples was performed in a similar fashion to the lettuce samples using both the AlphaLISA and the ELISA to validate the presence of Stx2 (Figure 4B). The resulting data was also similar with both the AlphaLISA and the ELISA demonstrating a significant difference between the 1× lysate dilution, the 10× lysate dilution and the control ($p < 0.0001$). However, the 100× lysate dilution could not be differentiated from the uninoculated control in either the AlphaLISA or the ELISA as $p = 0.18$ and 0.91, respectively. Taken together, the data from both the lettuce and the ground beef illustrate that the AlphaLISA is comparable to the ELISA in terms of sensitivity and the ability to detect the presence of Stx2 in inoculated food matrices.

2.4. AlphaLISA Sensitivity Using Increased Reagent Volumes

In an attempt to enhance the sensitivity of the AlphaLISA, a series of experiments were conducted using increased reagent volumes for the AlphaLISA (Figure 5). The premise behind these experiments was that the small sample volume of the AlphaLISA (5 µL) could be the limiting factor with regards to the sensitivity of the overall assay. Therefore, by simply increasing all assay components, the limit of detection may also increase accordingly. The ability of the AlphaLISA to detect 0, 0.5, 1.0, 3.0, or 10.0 ng/mL of purified Stx2 in the presence of a food matrix (lettuce or ground beef) using either the original assay components (1×), doubling the assay components (2×), or tripling the assay components (3×) was determined. From this, it was shown that when the assay components were increased from 1× to 2× that a corresponding increase in AlphaLISA signal was also detected in both lettuce and ground beef. However, the results were not consistent when assay components were increased to 3×. When the components of the assay were tripled, the AlphaLISA signal was higher than that seen for the 1× but not the 2× amounts in the lettuce; whereas this level resulted in the lowest signal in ground beef when compared to both the 1× and 2× assays.

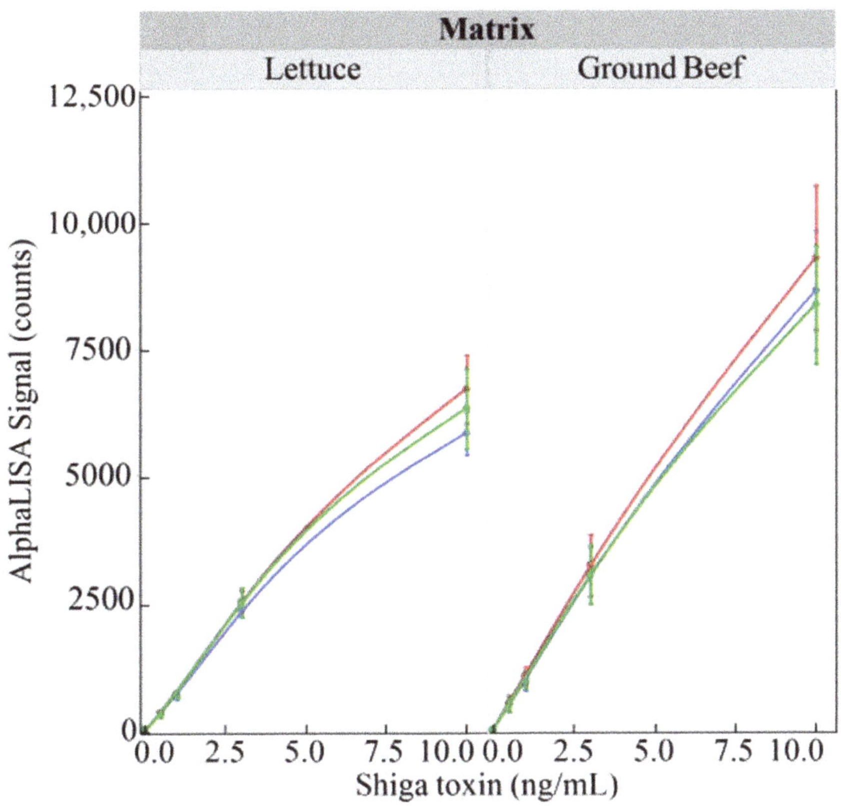

Figure 5. The effect of increasing reaction components on the sensitivity of the AlphaLISA for Stx2 in food matrices. Purified Stx2 (0, 0.5, 1.0, 3.0 or 5.0 ng/mL) was added to uninoculated lettuce (**left**) panel or ground beef (**right**) panel and the effect of increasing all of the reaction components (1×-blue line, 2×-red line, or 3×-green line) in the AlphaLISA signal was determined. Responses were quantified on a BioTek Cytation 5 with the emitted light from the AlphaLISA measured via Alpha mode at a gain setting of 100. Mean values from 2 independent trials, each containing 2 replicates are plotted with error bars denoting the standard deviation.

3. Discussion

AlphaLISA is a versatile technology that employs oxygen-channeling chemistry [27,28] and has been used for the detection of a wide variety of analytes from proteins to peptides to other small molecules. Currently, hundreds of different AlphaLISA biomarker detection kits are commercially available ranging in application from medicine (for tracing unwanted host cell proteins during industrial scale production of biotherapeutics), to agriculture (for the detection of aflatoxins in foods and animal feed), to basic research (as an alternative to the electrophoretic mobility shift assay for the detection of DNA-protein interactions). The AlphaLISA is considered a "no wash" alternative to the ELISA because it does not require any wash or separation steps [22]. The reagents are simply overlaid in succession, which greatly simplifies the protocol and makes the AlphaLISA highly amendable to high-throughput automated screening. Because of this, the AlphaLISA would be the superior assay when a considerable number of samples are to be analyzed under a standard operating procedure, such as is the case with the assessment of food quality and safety around the globe. Here, an AlphaLISA was developed to identify the presence of foodborne pathogens through the detection of Shiga toxin 2 in food samples.

Direct comparisons between the newly developed AlphaLISA and the ELISA were performed using both purified Stx2 and STEC-inoculated Romaine lettuce and ground beef samples. The suggestion provided by Perkin Elmer to estimate the limit of detection is to add 3 standard deviations to the mean of the "zero analyte" condition. Using this suggestion and the data presented in Figure 2, Table 1 was generated. Although the limits of detection were only slightly better for the AlphaLISA compared to the ELISA in terms of toxin concentration (0.10 ng/mL versus 0.14 ng/mL, respectively), when one considers that the ELISA employs a 100 μL sample while the AlphaLISA employs a 5 μL sample, quantitation of the amount of toxin on a per weight basis demonstrated that the AlphaLISA is the superior assay (0.5 pg for the AlphaLISA versus 14.3 pg for the ELISA). Interestingly, an attempt was made to increase the sensitivity of the AlphaLISA by doubling and tripling the components in the reaction including the amount of sample per assay (Figure 5). Although a corresponding increase in signal was detected when the reaction components were doubled (1× to 2× comparison), a tripling of the reaction components showed a decrease in signal intensity (2× to 3× comparison). The reason behind this deleterious effect with regard to signal production upon tripling of the reaction components is unknown and was not further pursued since the mean signal produced by the 1×, 2×, and 3× component mixtures were not statistically different.

Table 1. Comparisons of the limit of detection and signal-to-noise ratios of the AlphaLISA versus the ELISA.

Assay	Average Zero Signal	3× SD	LOD (ng/mL)	Signal-to-Noise (S:N) [a]			
				0.5 ng/mL	1 ng/mL	3 ng/mL	10 ng/mL
AlphaLISA	75.92	63.285	0.100	10	18	61	163
ELISA	0.21767	0.1665	0.143	4	7	18	17

SD = standard deviation, LOD = limit of detection. [a] Signal-to-noise ratios were calculated by dividing the signal mean of the stated assay upon the addition of either 0.5, 1, 3, or 10 ng/mL purified Stx2 to phosphate buffered saline (PBS) by the signal mean in samples with no Stx added to PBS.

The AlphaLISA is also superior because ELISAs typically have a narrower dynamic range, thus requiring the testing of multiple sample dilutions to accurately measure antigen concentration. As can be seen from Figure 3, samples within the range of 0.5 ng/mL to 3 ng/mL were detectable by the ELISA while samples containing 10 ng/mL of Stx2 fell outside of the absorbance range of the Cytation 5 microtiter plate reader, and thus could not be accurately measured unless diluted. Conversely, the AlphaLISA was able to measure concentrations within the 10 ng/mL range and possibly higher without dilution given the fact that AlphaLISA counts of ~1,600,000 have been recorded at higher gains using the Cytation 5 by Biotek (data not shown). Given that the saturation point was reached for ELISA but not for the AlphaLISA suggests that the dynamic range of the AlphaLISA is greater than that of the ELISA. While the observation of a broader dynamic range is consistent with what has been previously reported [22], larger scale studies would need to be conducted to precisely define the dynamic range of the assays. Regardless, the signal-to-noise ratio for the AlphaLISA was much larger than that of the ELISA (Table 1), indicating a better specification since there is more desired signal compared to background noise.

Antibody selection is crucial for the development of a functional AlphaLISA, including the fact that the specificity of the antibodies ultimately determines the accuracy of selection for the assay. The design presented here is specific for Stx2 (the more toxic of the two Stx types [11]) because it utilizes two different monoclonal antibodies with the donor antibody M-1005 specifically recognizing the Stx2 A-subunit and the acceptor M-1003 specifically recognizing the Stx2 B-subunit. Although the determination of which antibody was selected as the donor compared to the acceptor was done empirically, it has been hypothesized that this configuration was the most robust based on the following. First, the selection of two different antibodies ensures that the binding reaction can proceed without interference from the antibodies competing for binding sites, and allows the binding reaction to be more efficient compared to assays which utilize either the same antibody or antibodies that recognize different peptides within the same protein. Second, the AlphaLISA signaling cascade allows a single donor to activate multiple acceptors, and thus the creation of a more intense output signal compared to that of only a single acceptor in the presence of multiple donors. Because Stx2 consists of an A subunit noncovalently joined to a pentamer of identical B subunits, there are theoretically five times the number of acceptor particles surrounding every donor particle in the assay. Taken together, the configuration shown in Figure 1B would likely emit the most intense signal because the single A subunit donor could emit to multiple B subunit acceptors, which corresponded to the signal intensity observed experimentally. This result was analogous to what was seen for the detection of *Bacillus anthracis* spores via an AlphaLISA [29].

Ultimately, we developed a novel AlphaLISA for the detection of Stx2 and demonstrated its ability to identify the presence of Stx2 not only in phosphate buffered saline (PBS) but in two food matrices as well. Because the AlphaLISA can be performed in ~1.5 h (compared to the ~2.5 h needed for the ELISA), does not require the plate manipulation needed to wash the sample wells like the ELISA, and appears to have a larger dynamic range; the AlphaLISA is the superior method for the detection of Stx2 when large numbers of samples are to be queried. Other potential benefits may also be realized such as the fact that the AlphaLISA requires the toxin to be intact since the signal is dependent upon the A and B subunits being within a set distance to one another. This AlphaLISA might also be applicable for quantifying the amount of Stx2 in other matrices, indicating a use not only for detecting the presence of STECs in food as presented here but for use in clinical samples consisting of serum/stool [30,31] or for research investigating variables related to *stx* induction as well [32].

4. Materials and Methods

4.1. Preparation of Antibodies

Antibodies known to detect a majority of the Stx2 subtypes were selected from purified Abraxis (Warminster, PA, USA) in-house stocks stored in PBS (Abraxis LLC). The monoclonal antibody M-1003 binds the Shiga toxin B-subunits, while the monoclonal antibody M-1005 binds the Shiga toxin A-subunit. These two antibody stocks were also biotinylated using the Thermo Fisher Scientific (Waltham, MA, USA) EZ-Link Sulfo-NHS-LC-LC-Biotin (prod. #21338) according to manufacturer instructions, with a one-hour incubation at room temperature followed by removal of excess biotin with a Zeba Spin Desalting Column (Thermo Fisher Scientific prod. #89893). Biotinylated antibodies were stored in PBS at −20 °C. Antibody concentrations for both the biotinylated and stock antibodies were measured at 280 nm using a DeNovix microspectrophotometer (DeNovix Inc., Wilmington, DE, USA).

4.2. Preparation of AlphaLISA Donor and Acceptor Beads

Uncoupled AlphaLISA donor and acceptor beads were purchased from PerkinElmer, Inc. (Waltham, MA, USA; acceptor prod. #6772001 and donor prod. #6760002S). The streptavidin-coated donor beads did not need additional modification; however, the acceptor beads required antibody coupling prior to the start of the AlphaLISA assay. Two sets of acceptor beads were prepared (M-1003 and M-1005) using a modified version of the procedures found in the PerkinElmer "ELISA to Alpha Immunoassay Conversion Guide," section "Protocol for Direct Conjugation of an Antibody to an AlphaLISA Acceptor Bead" [33]. Briefly, two sets of 25 μL of 20 mg/mL acceptor beads (equaling 0.5 mg each) were washed with 25 μL of PBS in 1.5 mL microfuge tubes, centrifuged at 16,000× *g* for 15 min (Eppendorf centrifuge), and the supernatants discarded. To each set of 0.5 mg beads, 0.05 mg of either M-1003 or M-1005 antibody (non-biotinylated), 0.625 μL of 10% Tween-20 (Sigma-Aldrich, St. Louis, MO, USA; prod. #P7949), and 5 μL of 25 mg/mL $NaBH_3CN$ (Acros Organics, Thermo Fisher Scientific; prod #168550100) was added and brought to a final volume of 100 μL with 130 mM sodium phosphate (Sigma-Aldrich prod. #S0876/0). The bead pellet was resuspended in this mixture by pipetting, and incubated statically for 18 h at 37 °C. Blocking of unreacted sites was then performed by adding 5 μL of 65 mg/mL carboxy-methoxylamine (CMO) (Sigma-Aldrich prod. #C13408) in 0.8 M sodium hydroxide (Alfa Aesar, Haverhill, MA, USA; prod. #35631) to the previous reactions, and incubating for 1 h at 37 °C with an occasional gentle vortex. Finally, the beads underwent a series of wash steps as follows. The two conjugated, blocked bead solutions were centrifuged at 16,000× *g* for 15 min and the supernatants were discarded. The pellets were resuspended in 100 μL of 0.1 M Tris-HCl pH 8.0 (Sigma-Aldrich prod. #T3253) and centrifuged again at 16,000× *g* for 15 min. The previous step was repeated twice more, and the beads were resuspended by pipetting up and down in 100 μL of PBS with 0.05% ProClin-300 (Sigma-Aldrich prod. #48914) added as a preservative, for a final concentration of 5 mg/mL. The two separate antibody-conjugated acceptor bead solutions were vortexed and sonicated with twenty 1-s pulses in a water bath sonicator and stored at 4 °C in the dark.

4.3. Preparation of Shiga Toxin 2 Standards

Shiga toxin 2 was purchased from Toxin Technology (Sarasota, FL, USA, prod. STX-2), at 0.5 mg/mL, and diluted to 3, 30, and 300 ng/mL in AlphaLISA buffer (25 mM HEPES pH 7.4 (VWR International, Radnor, PA, USA; prod. #0511) containing the following from Sigma-Aldrich 0.5% Triton-X 100, 0.1% casein (prod. #C-5890), 1 mg/mL dextran (prod. #31390), and 0.05% ProClin-300prod. #48914).

4.4. AlphaLISA Antibody Pair and Concentration Optimization

Once the reagents (biotinylated antibodies, antibody-conjugated acceptor beads, and donor beads) and Shiga toxin 2 standards were prepared, AlphaLISA optimization began. Stx2 standards at 0, 3, 30, and 300 ng/mL in AlphaLISA buffer were evaluated to determine optimal antibody pairing and titer level. As per the recommended PerkinElmer procedures in the ELISA to Alpha Immunoassay Conversion Guide, 5 μL of Stx2 standards were added to 1/2-area plate (Perkin Elmer prod. #6005560) wells. Then, 10 μL of biotinylated M-1003 or M-1005 antibody (diluted to a final concentration of 0, 0.3, 1, or 3 nM in a 50 μL total reaction volume) was added, followed by 10 μL of M-1003 or M-1005 antibody-conjugated acceptor beads (diluted to a final concentration of 10 μg/mL in 50 μL total reaction volume). M-1003-conjugated acceptor beads were only added to wells that had biotinylated-M-1005 added, and M-1005-conjuated acceptor beads were only added to wells that had biotinylated-M-1003 added in order to have complementary antibody pairs instead of competing pairs. The plate was incubated at room temperature for 1 h. Next, 25 μL of streptavidin-coated donor beads were added (for a final concentration of 40 μg/mL in 50 μL total reaction volume) under low light conditions to prevent photobleaching of the beads. This results in a final 50 μL total reaction volume. The plate was incubated in the dark at room temperature for 30 min. Total in-assay workflow time was approximately 1 h and 45 min. Microtiter plates were then read using a BioTek Cytation 5 (BioTek Instruments, Inc., Winooski, VT, USA) in Alpha measurement mode with a gain setting of 100.

4.5. Analysis of Purified Shiga Toxin 2

Shiga toxin 2 standards were prepared as described above. Samples consisting of 5 μL of the 0, 0.5, 3, 30, and 300 ng/mL Stx2 standards were assayed via the AlphaLISA using only the 0.3 nM of biotinylated M-1005 antibody for association with the donor and the acceptor-conjugated M-1003 antibody as described above. ELISA testing was conducted simultaneously using the Abraxis Shiga Toxin 2 ELISA kit (Abraxis prod. #542010) using 100 μL of the 0, 0.5, 3, 30, and 300 ng/mL Stx2 standards following the manufacturer's instructions. Absorbencies for the ELISA were also measured on the BioTek Cytation 5 at 450 nm five min after the addition of stop solution.

4.6. Preparation of Lettuce and Ground Beef Samples

Lettuce and ground beef samples were prepared following the procedures described previously [15]. Briefly, Romaine lettuce was purchased from a local grocery store, the outermost leaves were removed, while the remaining leaves were chopped into ~1 cm^2 pieces on a 70% ethanol sterilized surface. Ground beef was also purchased from a local grocery store, separated into 25 g quantities, and kept frozen until used. STEC inoculum was prepared from a frozen *E. coli* O145:H28 bacterial stock by growing a small chip of frozen stock in Tryptic Soy Broth (TSB) (Becton Dickinson, East Rutherford, NJ, USA; prod. #211825) overnight at 37 °C shaking at 120 rpm in a 15 mL Corning Falcon tube. The following day, the overnight culture was serially diluted in glass dilution tubes with peptone water (Becton Dickinson prod. #218105), and stored at 4 °C. The number of colony forming units (CFU) present were determined by plating 1 mL of each dilution onto a Tryptic Soy Agar (TSA) plate (Becton Dickinson prod. #236950), which was incubated at 37 °C overnight. Inoculum levels were 9 and 10 CFU/mL for the lettuce and beef samples, respectively.

Samples were incubated in broth at a 1:4 ratio as follows. Lettuce or ground beef partitioned into 25 g samples was placed into filter Stomacher bags (Seward Laboratory Systems Inc., Islandia, NY, USA). One milliliter of ~10 CFU/mL STEC inoculum was added to the samples, or 1 mL of peptone water for the uninoculated negative control samples. Next, 75 mL of modified mTSB (TSB with 10 g/L Casamino Acids (Neogen Corp., Lansing, MI, USA; prod.# 7229A) and 66.7 ng/L mitomycin C (Sigma-Aldrich #M-0503) antibiotic toxin inducer), pre-warmed to 42 °C, was measured into all samples and hand-massaged to mix well and break up any clumps. Samples were subsequently incubated overnight (16 h) statically at 42 °C. The following day, 10 mL samples were removed from the outer portion of the Stomacher bags and stored at −80 °C in 15 mL Falcon tubes till use. Dilutions of lettuce and ground beef samples were made in AlphaLISA buffer in a similar fashion to that of the Shiga toxin 2 standards.

4.7. AlphaLISA and ELISA Comparison Food Matrices

Testing of the inoculated lettuce and ground beef samples described above was performed as follows. For the AlphaLISA, 5 μL of either inoculated or uninoculated control samples were added to Perkin Elmer 1/2-area plates, and tests were performed as detailed in the AlphaLISA optimization section above using the optimized conditions of 0.3 nM biotinylated-M-1005 antibodies with M-1003-conjugated acceptor beads. ELISA testing was conducted simultaneously using the Abraxis Shiga Toxin 2 ELISA kit using 100 μL of sample and following the manufacturer's instructions. Microtiter plates were read on a BioTek Cytation 5 for either the fluorescence signal using the Alpha mode (gain setting of 100) for the AlphaLISA or for absorbance at 450 nm five minutes after the addition of stop solution for the ELISA.

4.8. AlphaLISA Using 1×, 2× and 3× Components

Stx2 samples containing 0, 0.5, 1.0, 3.0, and 10.0 ng/mL of the Stx2 standards were prepared in uninoculated lettuce as described above. From there 5, 10 or 15 μL of the Stx2 containing lettuce was assayed using 10, 20, or 30 μL of prepared biotinylated antibody and 10, 20, or 30 μL of the prepared acceptor-conjugated beads for the 1×, 2×, and 3× experiments, respectively. (The AlphaLISA beads for all assays consisted of the 0.3 nM of biotinylated M-1005 antibody for association with the donor- and the acceptor-conjugated M-1003 antibody.) The plate was incubated at room temperature for 1 h before 25, 50, or 75 μL of streptavidin-coated donor beads were added under low light conditions resulting in total reaction volumes of 50 μL (1×), 100 μL (2×), or 150 μL (3×). Microtiter plates were incubated as before with the resulting signal quantified on the BioTek Cytation 5 as previously described. Identical assays were performed except uninoculated ground beef was used as the matrix instead of lettuce for these studies.

Author Contributions: A.G.G., C.M.A., F.M.R., J.A.C. and L.E.R. conceived and designed the experiments; A.G.G., L.E.R. and T.P.S.J. performed the experiments; A.G.G., C.M.A., J.A.C. and L.E.R. analyzed the data; F.M.R. contributed reagents/materials/analysis tools; C.M.A. and L.E.R. wrote the paper; and A.G.G., C.M.A., F.M.R., J.A.C., L.E.R. and T.P.S.J. edited the manuscript.

Funding: This material is based upon work supported by the U.S. Department of Agriculture, Agricultural Research Service, under Agreement No. 8072-42000-084. Mention of trade names or commercial products in this publication is solely for the purpose of providing specific information and does not imply recommendation or endorsement by the USDA. The USDA is an equal opportunity employer.

Conflicts of Interest: The authors declare no conflict of interest.

References

1. Griffin, P.M. *Escherichia coli* O157:H7 and other enterohemorrhagic *Escherichia coli*. In *Infections of the Gastrointestinal Tract*; Blaser, M.J., Smith, P.D., Ravdin, J.I., Greenberg, H.B., Guerrant, R.L., Eds.; Raven Press Ltd.: New York, NY, USA, 1995.
2. Scallan, E.; Hoekstra, R.M.; Angulo, F.J.; Tauxe, R.V.; Widdowson, M.A.; Roy, S.L.; Jones, J.L.; Griffin, P.M. Foodborne illness acquired in the United States-major pathogens. *Emerg. Infect. Dis.* **2011**, *17*, 7–15. [CrossRef] [PubMed]
3. Hoffmann, S.; Batz, M.B.; Morris, J.G. Annual cost of illness and quality-adjusted life year losses in the United States due to 14 foodborne pathogens. *J. Food Protect.* **2012**, *75*, 1292–1302. [CrossRef] [PubMed]
4. Smith, K.E.; Wilker, P.R.; Reiter, P.L.; Hedican, E.B.; Bender, J.B.; Hedberg, C.W. Antibiotic treatment of Escherichia coli O157 infection and the risk of hemolytic uremic syndrome, Minnesota. *Pediatr. Infect. Dis. J.* **2012**, *31*, 37–41. [CrossRef] [PubMed]
5. Wong, C.S.; Jelacic, S.; Habeeb, R.L.; Watkins, S.L.; Tarr, P.I. The risk of the hemolytic-uremic syndrome after antibiotic treatment of Escherichia coli O157:H7 infections. *New Engl. J. Med.* **2000**, *342*, 1930–1936. [CrossRef] [PubMed]
6. Tarr, P.I.; Gordon, C.A.; Chandler, W.L. Shiga-toxin-producing *Escherichia coli* and haemolytic uraemic syndrome. *Lancet* **2005**, *365*, 1073–1086. [CrossRef]
7. Allison, H.E. Stx-phages: Drivers and mediators of the evolution of STEC and STEC-like pathogens. *Future Microbiol.* **2007**, *2*, 165–174. [CrossRef] [PubMed]
8. Los, J.M.; Los, M.; Wegrzyn, G. Bacteriophages carrying Shiga toxin genes: Genomic variations, detection and potential treatment of pathogenic bacteria. *Future Microbiol.* **2011**, *6*, 909–924. [CrossRef] [PubMed]
9. Obrien, A.D.; Holmes, R.K. Shiga and Shiga-like toxins. *Microbiol. Rev.* **1987**, *51*, 206–220.
10. Law, D. Virulence factors of *Escherichia coli* O157 and other Shiga toxin-producing *E. coli*. *J. Appl. Microbiol.* **2000**, *88*, 729–745. [CrossRef] [PubMed]
11. Tesh, V.L.; Burris, J.A.; Owens, J.W.; Gordon, V.M.; Wadolkowski, E.A.; Obrien, A.D.; Samuel, J.E. Comparison of the relative toxicities of Shiga-like toxins type-I and type-II for mice. *Infect. Immun.* **1993**, *61*, 3392–3402. [PubMed]

12. Scheutz, F.; Teel, L.D.; Beutin, L.; Pierard, D.; Buvens, G.; Karch, H.; Mellmann, A.; Caprioli, A.; Tozzoli, R.; Morabito, S.; et al. Multicenter evaluation of a sequence-based protocol for subtyping Shiga toxins and standardizing Stx nomenclature. *J. Clin. Microbiol.* **2012**, *50*, 2951–2963. [CrossRef] [PubMed]
13. Anonymous. Detection and Isolation of Non-O157 Shiga Toxin-Producing *Escherichia coli* (STEC) from Meat Products and Carcass and Environmental Sponges. Available online: https://www.fsis.usda.gov/wps/wcm/connect/7ffc02b5-3d33-4a79-b50c-81f208893204/MLG-5B.pdf?MOD=AJPERES (accessed on 10 August 2018).
14. Anonymous. Detection, Isolation and Identification of *Escherichia coli* O157:H7 from Meat Products and Carcass and Environmental Sponges. Available online: https://www.fsis.usda.gov/wps/wcm/connect/51507fdb-dded-47f7-862d-ad80c3ee1738/MLG-5.pdf?MOD=AJPERES (accessed on 10 August 2018).
15. Gehring, A.G.; Fratamico, P.M.; Lee, J.; Ruth, L.E.; He, X.H.; He, Y.P.; Paoli, G.C.; Stanker, L.H.; Rubio, F.M. Evaluation of ELISA tests specific for Shiga toxin 1 and 2 in food and water samples. *Food Control* **2017**, *77*, 145–149. [CrossRef]
16. Kong, Q.; Patfield, S.; Skinner, C.; Stanker, L.H.; Gehring, A.; Fratamico, P.; Rubio, F.; Qi, W.; He, X. Validation of two new immunoassays for sensitive detection of a broad range of Shiga Toxins. *Austin Immunol.* **2016**, *1*, 1007.
17. Jung, B.Y.; Jung, S.C.; Kweon, C.H. Development of a rapid immunochromatographic strip for detection of *Escherichia coli* O157. *J. Food Protect.* **2005**, *68*, 2140–2143. [CrossRef]
18. Clotilde, L.M.; Bernard, C.T.; Hartman, G.L.; Lau, D.K.; Carter, J.M. Microbead-based immunoassay for simultaneous detection of Shiga toxins and isolation of *Escherichia coli* O157 in foods. *J. Food Prot.* **2011**, *74*, 373–379. [CrossRef] [PubMed]
19. DeCory, T.R.; Durst, R.A.; Zimmerman, S.J.; Garringer, L.A.; Paluca, G.; DeCory, H.H.; Montagna, R.A. Development of an immunomagnetic bead-immunoliposome fluorescence assay for rapid detection of *Escherichia coli* O157:H7 in aqueous samples and comparison of the assay with a standard microbiological method. *Appl. Environ. Microbiol.* **2005**, *71*, 1856–1864. [CrossRef] [PubMed]
20. Peppard, J.; Glickman, F.; He, Y.; Hu, S.I.; Doughty, J.; Goldberg, R. Development of a high-throughput screening assay for inhibitors of aggrecan cleavage using luminescent oxygen channeling (AlphaScreen (TM)). *J. Biomol. Screen* **2003**, *8*, 149–156. [CrossRef] [PubMed]
21. Roby, P.; Bosse, R.; Arcand, M. Multiplex Assay Methods and Compositions. U.S. Patent 8,486,719, 16 July 2013.
22. Eglen, R.M.; Reisine, T.; Roby, P.; Rouleau, N.; Illy, C.; Bosse, R.; Bielefeld, M. The use of AlphaScreen technology in HTS: Current status. *Curr. Chem. Genom.* **2008**, *1*, 2–10. [CrossRef] [PubMed]
23. Luna-Gierke, R.E.; Griffin, P.M.; Gould, L.H.; Herman, K.; Bopp, C.A.; Strockbine, N.; Mody, R.K. Outbreaks of non-O157 Shiga toxin-producing Escherichia coli infection: USA. *Epidemiol. Infect.* **2014**, *142*, 2270–2280. [CrossRef] [PubMed]
24. Robbins, A.; Anand, M.; Nicholas, D.C.; Egan, J.S.; Musser, K.A.; Giguere, S.; Prince, H.; Beaufait, H.E.; Sears, S.D.; Borda, J.; et al. Ground beef recall associated with Non-O157 Shiga toxin-producing Escherichia coli, United States. *Emerg. Infect. Dis.* **2014**, *20*, 165–167. [CrossRef] [PubMed]
25. Slayton, R.B.; Turabelidze, G.; Bennett, S.D.; Schwensohn, C.A.; Yaffee, A.Q.; Khan, F.; Butler, C.; Trees, E.; Ayers, T.L.; Davis, M.L.; et al. Outbreak of Shiga toxin-producing Escherichia coli (STEC) O157:H7 associated with romaine lettuce consumption, 2011. *PLoS ONE* **2013**, *8*, e55300. [CrossRef] [PubMed]
26. White, A.; Cronquist, A.; Bedrick, E.J.; Scallan, E. Food source prediction of Shiga toxin-producing Escherichia coli outbreaks using demographic and outbreak characteristics, United States, 1998–2014. *Foodborne Pathog. Dis.* **2016**, *13*, 527–534. [CrossRef] [PubMed]
27. Ullman, E.F.; Kirakossian, H.; Singh, S.; Wu, Z.P.; Irvin, B.R.; Pease, J.S.; Switchenko, A.C.; Irvine, J.D.; Dafforn, A.; Skold, C.N.; et al. Luminescent oxygen channeling immunoassay—Measurement of particle binding-kinetics by chemiluminescence. *Proc. Natl. Acad. Sci. USA* **1994**, *91*, 5426–5430. [CrossRef] [PubMed]
28. Ullman, E.F.; Kirakossian, H.; Switchenko, A.C.; Ishkanian, J.; Ericson, M.; Wartchow, C.A.; Pirio, M.; Pease, J.; Irvin, B.R.; Singh, S.; et al. Luminescent oxygen channeling assay (LOCI(TM)): Sensitive, broadly applicable homogeneous immunoassay method. *Clin. Chem.* **1996**, *42*, 1518–1526. [PubMed]
29. Mechaly, A.; Cohen, N.; Weiss, S.; Zahavy, E. A novel homogeneous immunoassay for anthrax detection based on the AlphaLISA method: Detection of *B. anthracis* spores and protective antigen (PA) in complex samples. *Anal. Bioanal. Chem.* **2013**, *405*, 3965–3972. [CrossRef] [PubMed]

30. Gould, L.H.; Bopp, C.; Strockbine, N.; Atkinson, R.; Baselski, V.; Body, B.; Carey, R.; Crandall, C.; Hurd, S.; Kaplan, R.; et al. Recommendations for diagnosis of shiga toxin—Producing *Escherichia coli* infections by clinical laboratories. *MMWR Recomm. Rep.* **2009**, *58*, 1–14. [CrossRef] [PubMed]
31. He, X.; Ardissino, G.; Patfield, S.; Cheng, L.W.; Silva, C.J.; Brigotti, M. An improved method for the sensitive detection of Shiga Toxin 2 in human serum. *Toxins* **2018**, *10*, 59. [CrossRef] [PubMed]
32. Marques, L.R.; Moore, M.A.; Wells, J.G.; Wachsmuth, I.K.; O'Brien, A.D. Production of Shiga-like toxin by *Escherichia coli*. *J. Infect. Dis.* **1986**, *154*, 338–341. [CrossRef] [PubMed]
33. PerkinElmer. ELISA to Alpha Immunoassay Conversion Guide. Available online: www.perkinelmer.com/lab-solutions/resources/docs/GDE_ELISAtoAlphaLISA.pdf (accessed on 7 September 2018).

toxins

MDPI

Article

Tissue Distribution and Elimination of Ciguatoxins in *Tridacna maxima* (*Tridacnidae*, Bivalvia) Fed *Gambierdiscus polynesiensis*

Mélanie Roué [1,*,†], Hélène Taiana Darius [2,†], André Ung [2], Jérôme Viallon [2], Manoella Sibat [3], Philipp Hess [3], Zouher Amzil [3] and Mireille Chinain [2]

1 Institut de Recherche pour le Développement (IRD)—UMR 241 EIO, PO box 53267, 98716 Pirae, Tahiti, French Polynesia

2 Institut Louis Malardé (ILM), Laboratory of Toxic Microalgae—UMR 241-EIO, PO Box 30, 98713 Papeete, Tahiti, French Polynesia; tdarius@ilm.pf (H.T.D.); aung@ilm.pf (A.U.); jviallon@ilm.pf (J.V.); mchinain@ilm.pf (M.C.)

3 IFREMER, Phycotoxins Laboratory, F-44311 Nantes CEDEX, France; manoella.sibat@ifremer.fr (M.S.); philipp.hess@ifremer.fr (P.H.); zouher.amzil@ifremer.fr (Z.A.)

* Correspondence: melanie.roue@ird.fr; Tel.: +689-40-416-413

† These authors contributed equally to this work.

Received: 17 April 2018; Accepted: 7 May 2018; Published: 10 May 2018

Abstract: Ciguatera is a foodborne disease caused by the consumption of seafood contaminated with ciguatoxins (CTXs). Ciguatera-like poisoning events involving giant clams (*Tridacna maxima*) are reported occasionally from Pacific islands communities. The present study aimed at providing insights into CTXs tissue distribution and detoxification rate in giant clams exposed to toxic cells of *Gambierdiscus polynesiensis*, in the framework of seafood safety assessment. In a first experiment, three groups of tissue (viscera, flesh and mantle) were dissected from exposed individuals, and analyzed for their toxicity using the neuroblastoma cell-based assay (CBA-N2a) and liquid chromatography-tandem mass spectrometry (LC-MS/MS) analyses. The viscera, flesh, and mantle were shown to retain 65%, 25%, and 10% of the total toxin burden, respectively. All tissues reached levels above the safety limit recommended for human consumption, suggesting that evisceration alone, a practice widely used among local populations, is not enough to ensure seafood safety. In a second experiment, the toxin content in contaminated giant clams was followed at different time points (0, 2, 4, and 6 days post-exposure). Observations suggest that no toxin elimination is visible in *T. maxima* throughout 6 days of detoxification.

Keywords: giant clams; ex situ exposure to toxic algae; ciguatoxins; *Gambierdiscus polynesiensis*; anatomical distribution; toxin elimination; CBA-N2a; LC-MS/MS

Key Contribution: Giant clams were experimentally exposed to highly toxic cells of *G. polynesiensis*. Ciguatoxins were preferentially accumulated in viscera, although toxin contents monitored in two other tissues, i.e. mantle and flesh, also exceeded the recommended safety limit. No toxin elimination was observed over a period of 6 days.

1. Introduction

Pacific Island Countries and Territories (PICTs) communities are strongly dependent on marine products for their subsistence and income [1,2]. Thus, seafood contamination by marine biotoxins represents a serious impediment to the economy of PICTs populations. Although fish are the most widely consumed marine products in PICTs, several species of marine invertebrates, including giant clams, are also part of the diet of local populations. As an example, reported catches of giant clam meat reached tens of tons annually during the 1970s and 1980s in Tonga, Fiji, and French Polynesia [3]. While similar levels of exploitation were still reported in the early 2000s from small and isolated islands of French Polynesia [4], the local market of giant clam meat from the East Tuamotu and Australes archipelagos now exceeds 70 tons per year [5]. Giant clams grow in quite shallow depths, so are easily harvested by hand. Even though some people still consume it, the hepatopancreas is most often eviscerated before consumption in local communities, probably due to its visual appearance and bitter taste, but also as a precaution (in the same way as the liver of fish is typically removed).

Ciguatera fish poisoning (CFP) is a seafood-borne illness classically caused by the consumption of tropical coral reef fish contaminated with ciguatoxins (CTXs), polyether neurotoxins produced by dinoflagellates of the genus *Gambierdiscus* [6]. The clinical manifestations of the disease classically involve gastrointestinal disorders, cold allodynia, itching, paresthesia, asthenia, muscular pain, and dizziness, together with other neurological and cardio-vascular symptoms [7]. Although coral reef fish are usually regarded as the primary vectors of ciguatera, atypical ciguatera-like poisoning events following the consumption of giant clams (*Tridacna maxima*), marine gastropods (*Tectus niloticus*), and sea urchins (*Tripneustes gratilla*) are also occasionally reported from PICTs, most notably from French Polynesia, New Caledonia, Cook Islands, and the Republic of Vanuatu [8–14]. Besides the symptoms typical of CFP, atypical signs were also observed, such as the rapid onset of the disease (occurring within minutes), alteration of taste, burning sensation on the tongue and throat, as well as paralysis. In the case of poisoning events involving *T. maxima*, toxicological analyses using the receptor binding assay (RBA) and the neuroblastoma cell-based assay (CBA-N2a) confirmed the presence of liposoluble compounds acting on voltage-gated sodium channels (VGSCs). These toxins also displayed a mode of action and chemical properties very similar to those of CTXs [11,15,16]. Unfortunately, confirmation of the exact nature of these compounds could not be achieved from wild samples.

Recently, Roué et al. (2016) [17] provided confirmation that giant clams should be regarded as potential vectors of ciguatera, particularly in fishing areas frequently exposed to *Gambierdiscus* blooms. These authors showed that *T. maxima* fed highly toxic cells of *Gambierdiscus polynesiensis* were able to bio-accumulate algal CTXs, at concentrations well above the safety limit commonly recommended for human consumption in the Pacific region. However, since toxin analyses were conducted on extracts prepared from whole specimens, it was not possible to tell whether CTXs were differentially concentrated in the tissues of contaminated animals. Likewise, the study by Roué et al. (2016) [17] did not provide any information about the detoxification time for CTXs in giant clams. Indeed, such information has significant public health implications and may help identify practical measures to limit the risk of poisoning among consumers, e.g., by simply removing the contaminated tissues and/or issuing a quarantine period within distribution channels whenever necessary.

In this context, two additional series of ex situ exposures to toxic *Gambierdiscus* cells were conducted to address the following issues: (i) assess the differential tissue distribution of CTXs in contaminated individuals; and (ii) assess whether any elimination of CTXs is visible from contaminated giant clams over a short-term period of one week, assuming that these bivalves belong to the group of "rapid detoxifiers" [18]. To achieve this, giant clams were exposed to in vitro cultures of the highly toxic strain TB92—*G. polynesiensis* in two independent experiments. In the first experiment ("tissue distribution study"), the exposed specimens were dissected into three tissues: viscera, mantle and flesh (i.e., rest of the body tissues, including gills) which were further tested for their toxicity. In the second experiment ("detoxification study"), contaminated individuals were sacrificed at different time intervals (0, 2, 4, and 6 days) following their transfer into clean water, and tested for their toxicity. Detection of CTXs in samples was achieved using the CBA-N2a and liquid chromatography—tandem mass spectrometry (LC-MS/MS) analyses.

2. Results

2.1. Tissue Distribution Study

Giant clams (n = 3 per tank) were placed in four independent tanks and exposed to *G. polynesiensis* cells (tank No. 1, 2, and 3) or kept in clean seawater (control tank). After two days of exposure, giant clams were sacrificed and dissected into three tissues: viscera (consisting mainly of the hepatopancreas), mantle and flesh (i.e., rest of the body tissues, including gills). Then, each group of tissues was pooled for each tank before further analyses.

Extracts obtained from viscera, mantle and flesh of control giant clams showed no toxicity as confirmed by CBA-N2a and LC-MS/MS analyses (data not shown).

Conversely, all extracts corresponding to viscera (n = 3), mantle (n = 3) and flesh (n = 3) of exposed giant clams were found toxic using CBA-N2a. As shown in Figure 1, corresponding to the mean of the three tanks for each tissue, a sigmoidal dose-response curve was obtained in OV^+ conditions (i.e., in the presence of ouabain and veratridine) whereas no cytotoxic effects were observed on Neuro-2a cells in OV^- conditions (i.e., in absence of ouabain and veratridine), a response typical of the presence of CTXs in these samples. The mean EC_{50} values (n = 3 tanks) were 0.24 ± 0.02, 1.21 ± 0.29, and 1.65 ± 0.52 μg tissue/μL for viscera, flesh and mantle, respectively (Figure 1).

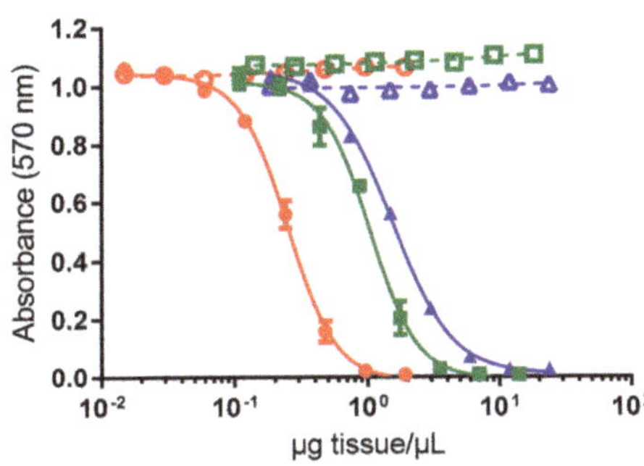

Figure 1. Dose-response curves of Neuro-2a cells in absence (open symbols) and presence (solid symbols) of ouabain and veratridine (OV^- and OV^+ conditions, respectively), when exposed to different concentrations of extracts obtained from viscera (○/●), flesh (□/■) and mantle (△/▲) of giant clams exposed to *G. polynesiensis* cells (TB92 strain). Data represent the mean ± SD of the three experimental tanks, each tissue tested in three independent CBA-N2a experiments (each run in triplicates).

Absolute toxicities of tissues (i.e., toxin contents in each tissue) from giant clams from each of the three tanks indicated that viscera was 4- to 6-fold more toxic than flesh and 4- to 9-fold more toxic than mantle (Table 1), with mean absolute toxicities (n = 3 tanks) of 13.0 ± 1.1, 2.7 ± 0.6, and 2.0 ± 0.6 ng P-CTX3C equiv./g tissue, respectively.

Regarding the relative weight contribution of each tissue to the whole body weight of giant clams, flesh was the most abundant tissue (around 50%), followed by mantle and viscera (around 25% each) (Table 1). The tissue distribution of toxins (as expressed per g of whole body) was then calculated as a function of absolute toxicity and relative weight contribution (Table 1). For each of the three tanks, despite the limited contribution of viscera to the total body mass of giant clams, the toxin distribution in this tissue was 2- to 3-fold and 5- to 8-fold higher than what is observed in the flesh and the mantle, respectively (Table 1). With a mean toxin distribution (n = 3 tanks) of 3.3 ± 0.3, 1.3 ± 0.3 and 0.5 ± 0.1 ng P-CTX3C equiv./g whole body for viscera, flesh and mantle, respectively (Figure 2a), the contribution of viscera to the whole toxin content averaged 65%, followed by the flesh and the mantle with 25% and 10%, respectively (Figure 2b).

Table 1. Absolute toxicities and tissue distribution of toxins in giant clams exposed to *G. polynesiensis* cells (TB92 strain) based on CBA-N2a data. In each tank, each group of tissues was pooled from three giant clams.

Tank	Tissue	Absolute Toxicity of Tissue [1]	Tissue Wet Weight [g] (Relative Weight Contribution, %)	Tissue Distribution of Toxins [2]	Tissue Contribution to the Total Body Toxin Content
1	Viscera	12.2 ± 0.4	36 (26%)	3.2 ± 0.1	59%
	Flesh	3.0 ± 0.2	75 (54%)	1.6 ± 0.1	31%
	Mantle	2.6 ± 0.2	28.5 (20%)	0.5 ± 0.0	10%
2	Viscera	13.1 ± 0.8	37.5 (25%)	3.3 ± 0.2	71%
	Flesh	2.0 ± 0.2	72 (48%)	0.9 ± 0.1	20%
	Mantle	1.4 ± 0.2	39.5 (27%)	0.4 ± 0.1	9%
3	Viscera	13.6 ± 1.4	33.5 (26%)	3.5 ± 0.4	64%
	Flesh	3.0 ± 0.1	58.5 (45%)	1.4 ± 0.0	25%
	Mantle	2.1 ± 0.3	38 (29%)	0.6 ± 0.1	11%

[1] The absolute toxicity of each tissue, expressed in ng P-CTX3C equiv./g tissue, was measured directly from the extracts of each tissue using the following formula: P-CTX3C EC_{50}/sample EC_{50}; [2] The tissue distribution of toxins, expressed in ng P-CTX3C equiv./g whole body, was calculated using the following formula: absolute toxicity × relative weight contribution.

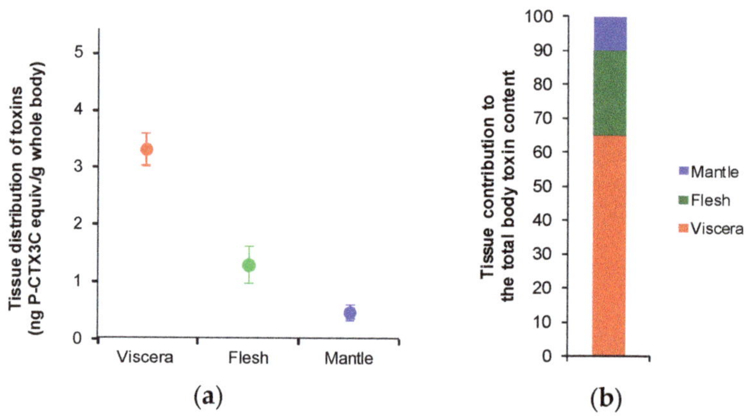

Figure 2. (**a**) Tissue distribution of toxins and (**b**) tissue contribution to the total body toxin content in giant clams exposed to *G. polynesiensis* cells (TB92 strain), as assessed by CBA-N2a data. Data represent the mean ± SD of the three experimental tanks, each tissue tested in three independent CBA-N2a experiments (each run in triplicates).

The presence of CTXs in toxic fractions was further confirmed by LC-MS/MS analyses, with P-CTX3B (RT 11.2 min) and P-CTX3C (RT 11.5 min) present in similar proportions (≈1:1) in all tissues (i.e., mantle, viscera or flesh) (Figure 3). The toxin content of viscera was higher than in the flesh and the mantle (where CTXs were detected at the limit of quantification) (Figure 3), confirming that viscera was the main storage tissue for CTXs compared to the other two tissue types. It should be noted that LC-MS/MS data in this study are mostly semi-quantitative as concentrations were low and no certified standards exist for any of the ciguatoxins.

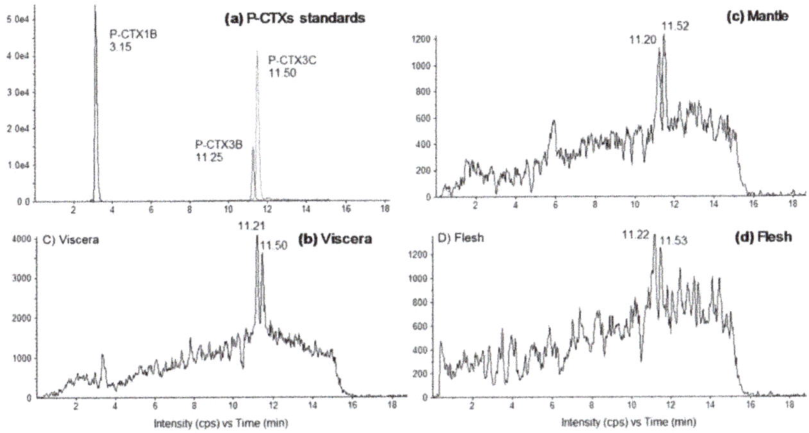

Figure 3. Liquid chromatography-tandem mass spectrometry (LC-MS/MS) chromatograms of (**a**) P-CTXs standards, and fractions obtained from (**b**) viscera, (**c**) mantle and (**d**) flesh of giant clams exposed to *G. polynesiensis* cells (TB92 strain). Chromatograms were acquired following the procedure described in Section 4.5, in positive multi-reaction monitoring mode. For (**d**), S/N < 3.

2.2. Detoxification Study

As in the first experiment, giant clams ($n = 3$ per tank) were placed in five independent tanks and exposed to *G. polynesiensis* cells ($n = 4$ tanks) or kept in clean seawater (control tank). After two days of exposure to *G. polynesiensis* cells, the seawater was renewed for the detoxification study. Each exposed tank was then randomly assigned a time point (0, 2, 4, and 6 days post-exposure) at which point the giant clams were sacrificed. After sacrifice, each whole animal was independently analyzed for its toxicity using CBA-N2a.

Fractions obtained from control animals showed no toxicity using CBA-N2a (data not shown). Conversely, all fractions obtained from contaminated giant clams after 0 ($n = 3$), 2 ($n = 3$), 4 ($n = 3$), and 6 ($n = 3$) days of detoxification were found toxic, with a sigmoidal dose-response curve typical of the presence of CTXs (data not shown). The mean toxin contents ($n = 3$ giant clams) were 1.8 ± 0.4, 2.5 ± 0.7, 2.0 ± 1.1, and 2.1 ± 0.7 ng P-CTX3C equiv./g whole body for 0, 2, 4, and 6 days of detoxification, respectively (Table 2). Thus, no significant elimination of toxins had occurred (ANOVA, $p > 0.05$) and whole giant clams contained a mean of 114 ± 54 ng P-CTX-3C equiv./giant clam all along the course of the detoxification period (Figure 4).

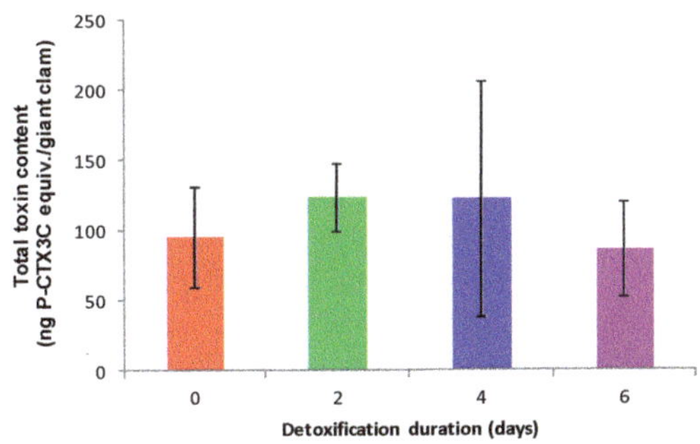

Figure 4. Mean total toxin burdens in contaminated giant clams, as assessed by CBA-N2a, at 0, 2, 4, and 6 days post-exposure. Data represent the mean ± SD of three individual giant clams, each of them tested in three independent CBA-N2a experiments (each run in triplicates).

Table 2. Toxin contents in contaminated giant clams after 0, 2, 4, and 6 days of detoxification, based on CBA-N2a data.

Days of Detoxification	Giant Clam Individual	Wet Weight [g]	Toxin Content [1]
0	1	39.9	1.4 ± 0.2
	2	57.1	1.9 ± 0.1
	3	60.8	2.2 ± 0.4
	Average of 1–3	53 ± 11	1.8 ± 0.4
2	1	39.0	3.3 ± 0.4
	2	40.8	2.3 ± 0.1
	3	67.2	1.9 ± 0.4
	Average of 1–3	49 ± 16	2.5 ± 0.7
4	1	71.0	3.3 ± 0.4
	2	54.3	1.6 ± 0.2
	3	57.6	1.1 ± 0.2
	Average of 1–3	61 ± 9	2.0 ± 1.1
6	1	46.6	1.4 ± 0.1
	2	43.9	2.8 ± 0.8
	3	32.1	2.0 ± 0.3
	Average of 1–3	41 ± 8	2.1 ± 0.7

[1] The toxin content, expressed in ng P-CTX3C equiv./g whole body, was measured directly from the extracts of each whole individual using the following formula: P-CTX3C EC_{50}/sample EC_{50}.

Each tank of three giant clams exposed to *G. polynesiensis* cells received a total toxin load of 17.25 µg P-CTX3C equiv. As shown in Figure 4, at the end of the exposure period (i.e., 0 days of detoxification), each giant clam had accumulated around 95 ng P-CTX3C equiv., consequently, it was estimated that each giant clam had retained only 0.6% of the total toxin load administered in each tank. This percentage stayed stable throughout the full detoxification experiment.

3. Discussion

3.1. Anatomical Distribution of CTXs in Toxic Giant Clams

Results of CBA-N2a analyses indicated that all the tissues (i.e., viscera, mantle, and rest of the flesh) of giant clams fed toxic *G. polynesiensis* cells, were able to retain CTXs. These findings confirmed previous observations on the ability of giant clams to bio-accumulate CTXs in their tissues [17]. In the present study, CTXs were preferentially concentrated in the viscera (mostly composed by the hepatopancreas), which was 2.5 and 6.7-fold more toxic than the flesh (i.e., the rest of body tissues, including gills) and mantle, respectively. This preferential accumulation of toxins in the viscera is consistent with a digestive uptake route of CTXs in this bivalve mollusk, as previously hypothesized by Roué et al. (2016) [17]. Indeed, following the filtration and ingestion of *G. polynesiensis* cells by *T. maxima*, the subsequent distribution of CTXs to non-visceral tissues could well explain the lower toxicities observed in the flesh and mantle. This differential accumulation of toxins in *T. maxima* tissues is coherent with the chemical nature of CTXs since lipophilic toxins are often concentrated in digestive glands; for instance, CTXs are known to be preferentially accumulated in the liver of fish [19,20]. Our results are also in agreement with previous studies conducted on other bivalve species, which concluded that in filter-feeding marine organisms exposed to harmful algal blooms, most of the toxin uptake accumulates in the viscera (most notably the digestive gland) despite the limited contribution of this tissue to the animal total body mass [18,21–30]. As an example, in the Japanese scallop *Patinopecten yessoensis* exposed to cultured *Dinophysis fortii* cells, dinophysistoxins (DTXs) and pectenotoxins (PTXs) were almost exclusively found in the digestive gland with only low levels being detected in the gill, mantle, gonad, and adductor muscle [27]. However, the preferential accumulation of toxins in non-digestive and non-visceral tissues has also been documented in a limited number of studies [28,31–33]. For example, the highest concentrations of okadaic acid-group (OA) toxins were detected in the adductor muscle of *M. chilensis* collected in areas where *Alexandrium catenella*, *Dinophysis acuminata* and *Dinophysis acuta* cells were observed [28]. Moreover, even if up to 90% of the azaspiracid (AZA) burden was found in digestive gland of mussels exposed to *Azadinium spinosum* cells, Jauffrais et al. (2012) [30] also showed that different algal densities or dissolved toxins may contribute to accumulation via different mechanisms, e.g., direct accumulation in gills or other tissues. Thus, the fact that up to approximately 40% of the total body burden was found in non-visceral tissues in this study may indicate that direct accumulation in mantle may also take place in giant clams.

In terms of public health implications, the distribution of toxins found in the viscera of toxic giant clams was 167-fold higher than the safety limit commonly recommended for human consumption, i.e., 0.01 ng P-CTX1B/g or 0.02 ng P-CTX3C/g [34,35]. Likewise, the flesh and mantle in toxic individuals also contained toxin levels well above this safety limit, i.e., 66 and 25-fold higher, respectively. Such observations clearly show that the common practice of removing the animal hepatopancreas before consumption of giant clams, will certainly contribute to reduce the amount of toxins ingested, but does not guarantee an effective protection of consumers as ca. 30–40% of the total toxin burden remains in non-visceral tissues. Similar conclusions apply to two bivalve mollusks, *V. antiqua* and *G. solida*, for which only the foot is consumed. Indeed, in their study, García et al. (2015) [28] showed that in addition to the digestive gland, foot also accumulates significant levels of toxins in these two species, and could thus be highly harmful to consumers. In contrast, the amount of DTXs detected in the adductor muscle of the Japanese scallops *P. yessoensis* was extremely low, confirming the validity of the practice of evisceration implemented in Japan since 1980 [27]. Still, poor dissection practices may also lead to contamination of non-visceral tissues as shown for domoic acid in scallops [36] and therefore, great care should be taken with evisceration as a means of toxin-reduction.

The TB92—*G. polynesiensis* strain is known to produce multiple P-CTX congeners, i.e., P-CTX3B, P-CTX3C, P-CTX3C/3B analogs, P-CTX4A, P-CTX4B, and M-seco-P-CTX4A [37], as confirmed by LC-MS/MS analyses performed on *Gambierdiscus* cultures used for exposure experiments [17]. Here, LC-MS/MS data indicated that P-CTX3B and P-CTX3C, present in almost similar proportions (assuming the same response factor in LC-MS/MS for P-CTX3B and P-CTX3C), were the only known analogs detected in all tissue samples, even though the potential presence of other CTX congeners at concentrations below the limit of detection of the technique cannot be completely ruled out. Interestingly, P-CTX3B and P-CTX3C were also among the major congeners detected in *Tectus niloticus* (Gastropod) and *Tripneustes gratilla* (Echinoid) specimens involved in CFP poisoning events in French Polynesia [13,14], suggesting that both congeners are preferentially bio-accumulated by marine invertebrates.

3.2. Elimination of CTXs from *T. maxima*

Results of CBA-N2a analyses showed that following the transfer of contaminated giant clams into clean seawater, the amount of toxins detected in *T. maxima* at 0, 2, 4, and 6 days post-exposure did not differ significantly. Moreover, giant clams were still not edible after one week in clean seawater (with a toxin content 105-fold higher than the safety limit commonly recommended for human consumption), suggesting that the time lime for toxin detoxification below the safety limit commonly recommended for human consumption in giant clams may take weeks or even months. These results are consistent with field studies conducted on other marine invertebrates species also involved in CFP poisoning events in French Polynesia, which indicated that *Tectus niloticus* (Gastropod) and *Tripneustes gratilla* (Echinoid) have a slow detoxification rate for CTXs. Indeed, based on CBA-N2a results, a 19-fold and 64-fold decrease in the overall ciguatoxicity of *Tectus niloticus* and *Tripneustes gratilla* samples, respectively, was observed over a two-year period, but toxin concentrations were still consistently above the safety limit recommended for human consumption [13,14]. Of note, the authors also underlined that these findings may be confounded by the additional accumulation of CTXs over the study period, even in the presence of low cell numbers of *Gambierdiscus*.

Elimination of toxins is mainly affected by their chemical properties. In general, lipophilic toxins are retained longer than the hydrophilic toxins. In bivalve mollusks, this elimination rate has been shown to also vary according to the species involved, taking from days up to several months, depending on initial concentration and environmental conditions [38]. Classically, bivalve species fall into two general categories in terms of their detoxification capacity: (i) rapid to moderate detoxifiers which eliminate toxins in a few days; and (ii) slow detoxifiers which need several months or even years to eliminate toxins [18,31,33,39–41]. For example, the blue mussel *Mytilus edulis* eliminate PSTs in less than 10 days whereas the butter clam *Saxidomus giganteus* need one to four months to depurate the same toxins [18]. In the same way, the clam *Perna viridis* showed a detoxification time for brevetoxins (PbTxs) significantly higher (above 4–5 months) than the oyster *Crassostrea virginica* and the clam *Mercenaria* after a bloom of *Karenia brevis* [41]. Finally, DA detoxification time has been shown to be species-specific and to have a wide-ranging variability, since most mussels depurate DA very quickly [39] whereas the razor clam *Siliqua patula* or the scallop *P. maximus* have much slower detoxification kinetics [31,40]. Results in the present study suggest that *T. maxima* falls into the slow detoxifiers' category since no significant toxin elimination was visible throughout one week of detoxification; however, further experiments with longer detoxification period will be necessary to firmly conclude this. Furthermore, as for other marine invertebrates [13,14], it would be interesting to study bioaccumulation and elimination rates of CTXs in wild giant clams.

At the end of the exposure period, each giant clam had retained only 0.6% of the toxin burden administered to each tank, corresponding to a total accumulation of 1.8% for each tank containing three giant clams, a result coherent with the study by Roué et al. (2016) [17] which showed an accumulation of around 3% of the toxins in the same experimental conditions. This low toxin uptake rate observed in giant clams is not surprising. For example, a study has shown that the absorption efficiency of DTX1 by the digestive gland of *P. yessoensis* was estimated at less than 3% of the total amount of DTX1 given to the scallops that were fed *D. fortii* cells [42]. Likewise, the bay scallops *Argopecten irradians* exposed to cultured cells of *Prorocentrum lima* showed a toxin assimilation efficiency in their tissues of less than 1% [43]. Interestingly, a biphasic detoxification kinetic is often observed in bivalves, consisting in an initial, more rapid, detoxification phase and a subsequent slower phase of toxin elimination [18,25,30,44,45]. It has been suggested that the initial rapid loss of toxin corresponds to clearance of unassimilated toxins from undigested cells or in dissolved form in the gut lumen, whereas the second slower phase represents the release of assimilated toxins either bound to the digestive gland or incorporated within other tissues [21,25,30,46,47]. In the present study, one hypothesis could be that giant clams could also have a biphasic detoxification kinetic, and that the rapid detoxification of unassimilated CTXs in contaminated animals actually occurred during the exposure period, and that these latter were in the slow phase of toxin elimination during the course of the detoxification study. Furthermore, another hypothesis could be that this low toxin uptake is in fact due to a low ingestion of *Gambierdiscus* cells by giant clams. In any case, further experiments with a longer exposure time may be necessary to better reflect conditions in the natural environment, and should be followed by a longer detoxification period and search for toxins in feces and system circulating-water as well as quantification of remaining *Gambierdiscus* cells in experimental tanks, in order to clarify actual accumulation and detoxification rates in giant clams.

4. Materials and Methods

4.1. Biological Material

4.1.1. Giant Clams

Giant clams (*T. maxima*) used in this study were purchased from an aquaculture farm in Tahiti (French Polynesia). For the tissue distribution study, giant clams (n = 12) were 15.3 ± 1.1 cm long, had a mean shell height of 8.4 ± 1.4 cm and provided a mean flesh wet weight of 60.6 ± 16.6 g. For the detoxification study, giant clams (n = 15) were 17.1 ± 1.6 cm long, had a mean shell height of 9.9 ± 1.4 cm and gave a mean flesh wet weight of 54.4 ± 14.2 g.

4.1.2. Cultures of *Gambierdiscus polynesiensis*

A highly toxic strain (TB92—*G. polynesiensis*) available from the algal collection of Institut Louis Malardé was used for the ex situ exposure experiments. TB92 cultures were obtained as previously described in Roué et al. (2016) [17]. Cultures were harvested in their late exponential/early stationary growth phase (i.e., 28-days post-inoculation) when cells exhibit highest CTX levels corresponding to an average toxicity of 5.83 ± 0.85 pg P-CTX3C equiv./cell.

4.2. Ex Situ Exposure Experiments

The experimental set-up was similar to the one previously described in Roué et al. (2016) [17]. Experiments were conducted in a closed environment, in tanks containing 20 L of seawater with a salinity value of 37. The temperature and the percentage of dissolved oxygen were stabilized at around 28.8 °C and 8 mg/L, respectively, and controlled daily. The light regime followed a 12:12 h (light:dark) photoperiod with an average irradiance of about 50 µmol photons/m/s of light (daylight fluorescent tubes). Each tank was equipped with a pump set at a flow rate of 200 L/h, in order to favor the suspension and/or dissemination of *Gambierdiscus* cells in the surrounding environment of giant clams. Three giant clams were placed in each experimental tank and acclimated to laboratory conditions during three days prior to exposure. Two independent series of experiments were conducted, in order to assess both the distribution of CTXs in three tissue types and the detoxification rate of CTXs from giant clams, respectively.

4.2.1. Tissue Distribution Study

In three tanks, a cell dose of 150,000 cells was administered 20 times to giant clams over a period of 48 h, to reach a total of 3×10^6 *Gambierdiscus* cells per tank, corresponding to an overall toxin load of 17.25 µg P-CTX3C equiv., which is representative of what can be found in natural blooms, as previously discussed in Roué et al. (2016) [17]. A fourth tank containing animals kept in the same laboratory conditions as exposed animals served as control. The water was not replaced in the course of the exposure period. No mortality was observed during the acclimation step, nor in the course of the exposure experiment.

At the end of the exposure period, each individual was collected and immediately sacrificed. The whole meat was removed from the shell and thoroughly rinsed twice in 0.5 L seawater to remove all trace of incubation water or intervalvar liquid potentially contaminated with *Gambierdiscus* cells and/or dissolved CTXs. Each giant clam sample was further dissected into three tissues: viscera (consisting mainly of the hepatopancreas), mantle and flesh (i.e., rest of the body tissues, including gills). Each group of tissue corresponding to the three giant clams from a same experimental tank was then pooled, ground and stored at −20 °C.

4.2.2. Detoxification Study

Four tanks received each a total cell load of 3×10^6 cells throughout a 48 h exposure period, corresponding to an overall toxin load of 17.25 µg P-CTX3C equiv., in the same conditions that for the tissue distribution study. A fifth tank containing animals maintained in the same laboratory conditions as exposed animals served as control. Following toxic exposure, seawater in each of the five tanks was entirely renewed by clean seawater in order to remove any remaining *Gambierdiscus* cells (data not available) and/or dissolved CTXs. In order, to assess whether any elimination of CTXs is visible from contaminated giant clams over a short-term period of one week, assuming that these bivalves belong to the group of "rapid detoxifiers" [18], each exposed tank was then randomly assigned a time point (0, 2, 4, and 6 days post-exposure) at which to sacrifice the three individuals from the respective tank. For each animal, the whole meat was removed from the shell and thoroughly rinsed twice in 0.5 L seawater. Each sample was then ground separately and stored at −20 °C.

4.3. Toxin Extration

Samples were extracted according to the procedure previously described in Roué et al. (2016) [17]. Briefly, 10 g of a given tissue (for tissue distribution study) or 10 g of the whole giant clam body (for detoxification study) were extracted in pure methanol (MeOH), followed by 50% aqueous MeOH. The resulting dried extract was partitioned between dichloromethane (CH_2Cl_2) and 60% aqueous MeOH. The CH_2Cl_2 phase, likely to contain CTXs, was dried under vacuum and defatted by a second solvent partition using cyclohexane and aqueous MeOH. The methanolic fraction was recovered and further purified using C_{18} Sep-Pak cartridges (360 mg, Waters®). The resulting 90% aqueous methanol fraction, likely to contain the majority of CTXs, was dried in a SpeedVac concentrator and stored at +4 °C until CBA-N2a and LC-MS/MS analyses.

4.4. Neuroblastoma Cell-Based Assay (CBA-N2a)

CBA-N2a analyses were conducted following the procedure previously described in Roué et al. (2016) [17] as adapted from Manger et al. (1993) [48]. All dry fractions were carefully weighed and re-suspended in methanol to reach a final concentration of 1 mg of dry extract/100 µL prior to CBA-N2a assays. Fractions were tested at a concentration range of 5–9524 pg/µL in eight distinct concentrations and three independent experiments (each run in triplicate). Absorbance data were fitted to a sigmoidal dose-response curve (variable slope) based on the four-parameter logistic model (4PL) allowing the calculation of EC_{50} values using Prism v6.0.7 software (GraphPad, San Diego, CA, USA). The half maximal effective concentration values (EC_{50}) $\pm$ standard deviations (SD) were expressed in g tissue equiv./µL or g whole body equiv./µL for tissue distribution and detoxification studies, respectively. Toxin contents (T) were then estimated using the following formula T = (P-CTX3C EC_{50}/sample EC_{50}) and expressed in ng P-CTX3C equiv./g tissue or ng P-CTX3C equiv./g whole body for tissue distribution and detoxification studies, respectively. According to Bricelj and Shumway (1998) [18], the contribution of a tissue to the total toxin body burden is a function of both its absolute toxicity and relative weight contribution. Thus, the tissue distribution of toxins, expressed in ng P-CTX3C equiv./g whole body, were calculated for each tissue (viscera, mantle, and flesh) using the following formula: absolute toxicity (expressed in ng P-CTX3C equiv./g tissue) $\times$ relative weight contribution (in %). The EC_{50} value obtained for P-CTX3C, the maximum concentration of dry extracts (MCE) to be tested in CBA-N2a for giant clam fractions and the limit of quantification (LOQ) for CBA-N2a in our assays conditions were also assessed and were 3.10 $\pm$ 0.76 fg/µL, 10,869 pg/µL, and 0.014 ng P-CTX3C equiv./g of giant clam, respectively.

4.5. Liquid Chromatography Coupled with Tandem Mass Spectrometry (LC-MS/MS)

The procedure for liquid chromatography coupled with tandem mass spectrometry (LC-MS/MS) analyses was adapted from Yogi et al. (2011) [49]. Analyses were performed using an LC system (UFLC XR Nexera, SHIMADZU, Kyoto, Japan) coupled to a hybrid triple quadrupole-linear ion trap API4000 QTRAP mass spectrometer (SCIEX, Redwood City, CA, USA) equipped with a TurboV® electrospray ionization source. A Zorbax Eclipse plus column (C_{18}, 1.8 µm, 50 mm $\times$ 2.1 mm, Agilent technologies, Santa Clara, CA, USA) was employed at 40 °C and P-CTXs were eluted at 400 µL/min with a linear gradient using water as eluent A and methanol as eluent B, both eluents containing 2 mM ammonium formate and 50 mM formic acid. The elution gradient ran from 78 to 88% B over 10 min and was held for 4 min before returning to initial conditions and re-equilibration during 5 min. Five microliters of sample were injected onto the column. The instrument control, data processing and analysis were conducted using Analyst software. The API4000 QTRAP was operated in positive mode using Multi Reaction Monitoring (MRM). The pseudomolecular ions $[M+NH_4]^+$ and $[M+H]^+$ were selected as precursor ions. The ions resulting in the successive losses of NH_4 and/or water molecules were selected as product ions (see Roué et al. (2016) [17]). The MRM experiments were established by using the following source settings: curtain gas set at 25, ion spray at 5500 V, a turbogas temperature of 300 °C, gas 1 set at 40 and gas 2 set at 60 psi with an entrance potential of 10 V. Calibration was carried out using P-CTX3C standard purchased from Wako chemical. In addition, three P-CTX standards: P-CTX1B, P-CTX3B, and P-CTX3C, obtained from the Institut Louis Malardé's bank of standards were injected to obtain a toxin profile for verification of retention times. The limit of detection (LOD) was statistically determined and found to be 0.66 ng P-CTX3C equiv./g of shellfish matrix.

Author Contributions: M.R., H.T.D. and M.C. conceived of and designed the study. A.U. performed the in vitro cultures of *Gambierdiscus* cells for exposure experiments. Mélanie Roué performed the tissue distribution and detoxification studies. Mélanie Roué performed chemical extractions prior to CBA-N2a and LC-MS/MS analyses. J.V. and M.S. conducted the CBA-N2a and LC-MS/MS analyses, respectively. M.R., H.T.D. and J.V. contributed to the CBA-N2a data analysis and M.S., P.H. and Z.A. to the LC-MS/MS data analysis. M.R. drafted the paper and coordinated its revisions while all authors contributed to the editing of the paper.

Acknowledgments: The work described in this paper was supported by funds from the countries of France and French Polynesia (Contrat de Projet "Caractérisation écotoxicologique des lagons exploités" CELEX—No. 8192-2012/VP/DBP). The authors greatly acknowledge Nicolas Maihota (IRD), Philippe Cruchet (ILM) and Laura Dufranc (IRD/ILM) for their technical assistance.

Conflicts of Interest: The authors declare no conflict of interest.

References

1. Bell, J.D.; Kronen, M.; Vunisea, A.; Nash, W.J.; Keeble, G.; Demmke, A.; Pontifex, S.; Andréfouët, S. Planning the use of fish for food security in the Pacific. *Mar. Policy* **2009**, *33*, 64–76. [CrossRef]
2. Charlton, K.E.; Russell, J.; Gorman, E.; Hanich, Q.; Delisle, A.; Campbell, B.; Bell, J. Fish, food security and health in Pacific Island countries and territories: A systematic literature review. *BMC Public Health* **2016**, *16*. [CrossRef] [PubMed]
3. Van Wynsberge, S.; Andréfouët, S.; Gaertner-Mazouni, N.; Wabnitz, C.C.C.; Gilbert, A.; Remoissenet, G.; Payri, C.; Fauvelot, C. Drivers of density for the exploited giant clam *Tridacna maxima*: A meta-analysis. *Fish Fish.* **2016**, *17*, 567–584. [CrossRef]
4. Gilbert, A.; Andréfouët, S.; Yan, L.; Remoissenet, G. The giant clam *Tridacna maxima* communities of three French Polynesia islands: Comparison of their population sizes and structures at early stages of their exploitation. *ICES J. Mar. Sci.* **2006**, *63*, 1573–1589. [CrossRef]
5. Direction des Ressources Marines et Minières de Polynésie Française. Available online: http://www.peche.pf/spip.php?rubrique213 (accessed on 27 February 2018).
6. Dickey, R.W.; Plakas, S.M. Ciguatera: A public health perspective. *Toxicon* **2010**, *56*, 123–136. [CrossRef] [PubMed]
7. Friedman, M.A.; Fernandez, M.; Backer, L.C.; Dickey, R.W.; Bernstein, J.; Schrank, K.; Kibler, S.; Stephan, W.; Gribble, M.O.; Bienfang, P.; et al. An updated review of ciguatera fish poisoning: Clinical, epidemiological, environmental, and public health management. *Mar. Drugs* **2017**, *15*, 72. [CrossRef] [PubMed]
8. Bagnis, R. A propos de quelques cas d'intoxications par des mollusques du genre "bénitier" dans une île de la Société. *Bull. Soc. Pathol. Exot.* **1967**, *60*, 580–592.
9. Bagnis, R. Situation de l'endémicité ciguatérique aux îles Gambier. *Cahiers Pacifique* **1974**, *18*, 585–600.
10. Rongo, T.; van Woesik, R. Socioeconomic consequences of ciguatera poisoning in Rarotonga, southern Cook Islands. *Harmful Algae* **2012**, *20*, 92–100. [CrossRef]
11. Laurent, D.; Kerbrat, A.S.; Darius, H.T.; Rossi, F.; Yeeting, B.; Haddad, M.; Golubic, S.; Pauillac, S.; Chinain, M. Ciguatera Shellfish Poisoniong (CSP): A new ecotoxicological phenomenon from cyanobacteria to humans via giant clams. In *Food Chains: New Research*; Jensen, M.A., Muller, D.W., Eds.; Nova Science Publishers, Inc.: New York, NY, USA, 2012; pp. 1–44. ISBN 9781613243770.
12. Gatti, C.; Lonati, D.; Darius, H.T.; Zancan, A.; Roué, M.; Schicchi, A.; Locatelli, C.A.; Chinain, M. *Tectus niloticus* (*Tegulidae*, Gastropod) as a Novel Vector of Ciguatera Poisoning: Clinical Characterization and Follow-Up of a Mass Poisoning Event in Nuku Hiva Island (French Polynesia). *Toxins* **2018**, *10*, 102. [CrossRef] [PubMed]
13. Darius, H.T.; Roué, M.; Sibat, M.; Viallon, J.; Gatti, C.M.I.; Vandersea, M.W.; Tester, P.A.; Litaker, R.W.; Amzil, Z.; Hess, P.; et al. *Tectus niloticus* (*Tegulidae*, Gastropod) as a Novel Vector of Ciguatera Poisoning: Detection of Pacific Ciguatoxins in Toxic Samples from Nuku Hiva Island (French Polynesia). *Toxins* **2018**, *10*, 2. [CrossRef] [PubMed]

14. Darius, H.T.; Roué, M.; Sibat, M.; Viallon, J.; Gatti, C.; Vandersea, M.; Tester, P.A.; Litaker, R.W.; Amzil, Z.; Hess, P.; et al. Toxicological investigations on the sea urchin *Tripneustes gratilla* (*Toxopneustidae*, Echinoid) from Anaho Bay (Nuku Hiva, French Polynesia): Evidence for the presence of Pacific ciguatoxins. *Mar. Drugs* **2018**, *16*, 122. [CrossRef] [PubMed]
15. Laurent, D.; Kerbrat, A.S.; Darius, H.T.; Girard, E.; Golubic, S.; Benoit, E.; Sauviat, M.P.; Chinain, M.; Molgo, J.; Pauillac, S. Are cyanobacteria involved in Ciguatera Fish Poisoning-like outbreaks in New Caledonia? *Harmful Algae* **2008**, *7*, 827–838. [CrossRef]
16. Pawlowiez, R.; Darius, H.T.; Cruchet, P.; Rossi, F.; Caillaud, A.; Laurent, D.; Chinain, M. Evaluation of seafood toxicity in the Australes archipelago (French Polynesia) using the neuroblastoma cell-based assay. *Food Addit. Contam. Part A Chem. Anal. Control. Expo. Risk Assess.* **2013**, *30*, 567–586. [CrossRef] [PubMed]
17. Roué, M.; Darius, H.T.; Picot, S.; Ung, A.; Viallon, J.; Gaertner-Mazouni, N.; Sibat, M.; Amzil, Z.; Chinain, M. Evidence of the bioaccumulation of ciguatoxins in giant clams (*Tridacna maxima*) exposed to *Gambierdiscus* spp. cells. *Harmful Algae* **2016**, *57*, 78–87. [CrossRef]
18. Bricelj, V.M.; Shumway, S.E. Paralytic Shellfish Toxins in Bivalve Molluscs: Occurrence, Transfer Kinetics, and Biotransformation. *Rev. Fish. Sci.* **1998**, *6*, 315–383. [CrossRef]
19. Lewis, R.J.; Sellin, M.; Street, R.; Holmes, M.H.; Gillepsie, N.C. Excretion of ciguatoxin from moray eels (Muraenidae) of the central Pacific. In *Proceedings of the Third International Conference on Ciguatera Fish Poisoning*; Tosteson, T.R., Ed.; Polysciences Publications: Quebec, QC, Canada, 1992; pp. 131–143.
20. Chateau-Degat, M.L.; Huin-Blondey, M.O.; Chinain, M.; Darius, T.; Legrand, A.M.; Ngoc, L.N.; Laudon, F.; Chansin, R.; Dewailly, E. Prevalence of chronic symptoms of ciguatera disease in French Polynesian adults. *Am. J. Trop. Med. Hyg.* **2007**, *77*, 842–846. [PubMed]
21. Chen, C.Y.; Chou, H.N. Accumulation and depuration of paralytic shellfish poisoning toxins by purple clam *Hiatula rostrata* Lighttoot. *Toxicon* **2001**, *39*, 1029–1034. [CrossRef]
22. Blanco, J.; Acosta, C.P.; Bermúdez de la Puente, M.; Salgado, C. Depuration and anatomical distribution of the amnesic shellfish poisoning (ASP) toxin domoic acid in the king scallop *Pecten maximus*. *Aquat. Toxicol.* **2002**, *60*, 111–121. [CrossRef]
23. Kwong, R.W.M.; Wang, W.X.; Lam, P.K.S.; Yu, P.K.N. The uptake, distribution and elimination of paralytic shellfish toxins in mussels and fish exposed to toxic dinoflagellates. *Aquat. Toxicol.* **2006**, *80*, 82–91. [CrossRef] [PubMed]
24. Blanco, J.; Mariño, C.; Martín, H.; Acosta, C.P. Anatomical distribution of diarrhetic shellfish poisoning (DSP) toxins in the mussel *Mytilus galloprovincialis*. *Toxicon* **2007**, *50*, 1011–1018. [CrossRef] [PubMed]
25. Lassus, P.; Amzil, Z.; Baron, R.; Séchet, V.; Barillé, L.; Abadie, E.; Bardouil, M.; Sibat, M.; Truquet, P.; Bérard, J.B.; et al. Modelling the accumulation of PSP toxins in Thau Lagoon oysters (*Crassostrea gigas*) from trials using mixed cultures of *Alexandrium catenella* and *Thalassiosira weissflogii*. *Aquat. Living Resour.* **2007**, *20*, 59–67. [CrossRef]
26. Murray, S.A.; O'Connor, W.A.; Alvin, A.; Mihali, T.K.; Kalaitzis, J.; Neilan, B.A. Differential accumulation of paralytic shellfish toxins from *Alexandrium minutum* in the pearl oyster, *Pinctada imbricata*. *Toxicon* **2009**, *54*, 217–223. [CrossRef] [PubMed]
27. Matsushima, R.; Uchida, H.; Nagai, S.; Watanabe, R.; Kamio, M.; Nagai, H.; Kaneniwa, M.; Suzuki, T. Assimilation, accumulation, and metabolism of dinophysistoxins (DTXs) and pectenotoxins (PTXs) in the several tissues of Japanese scallop *Patinopecten yessoensis*. *Toxins* **2015**, *7*, 5141–5154. [CrossRef] [PubMed]
28. García, C.; Pérez, F.; Contreras, C.; Figueroa, D.; Barriga, A.; López-Rivera, A.; Araneda, O.F.; Contreras, H.R. Saxitoxins and okadaic acid group: Accumulation and distribution in invertebrate marine vectors from Southern Chile. *Food Addit. Contam. Part A Chem. Anal. Control. Expo. Risk Assess.* **2015**, *32*, 984–1002. [CrossRef] [PubMed]
29. Kameneva, P.A.; Krasheninina, E.A.; Slobodskova, V.V.; Kukla, S.P.; Orlova, T.Y. Accumulation and tissue distribution of dinophysitoxin-1 and dinophysitoxin-3 in the mussel *Crenomytilus grayanus* feeding on the benthic dinoflagellate *Prorocentrum foraminosum*. *Mar. Drugs* **2017**, *15*, 330. [CrossRef] [PubMed]
30. Jauffrais, T.; Marcaillou, C.; Herrenknecht, C.; Truquet, P.; Séchet, V.; Nicolau, E.; Tillmann, U.; Hess, P. Azaspiracid accumulation, detoxification and biotransformation in blue mussels (*Mytilus edulis*) experimentally fed *Azadinium spinosum*. *Toxicon* **2012**, *60*, 582–595. [CrossRef] [PubMed]
31. Drum, A.S.; Siebens, T.L.; Crecelius, E.A.; Elston, R.A. Domoic acid in the Pacific razor clam *Siliqua patula* (Dixon, 1789). *J. Shellfish Res.* **1993**, *12*, 443–450.

32. Wekell, J.C.; Gauglitz, E.J.; Bamett, H.J.; Hatfield, C.L.; Simons, D.; Ayres, D. Occurrence of domoic acid in washington state razor clams (*Siliqua patula*) during 1991–1993. *Nat. Toxins* **1994**, *2*, 197–205. [CrossRef] [PubMed]
33. Sagou, R.; Amanhir, R.; Taleb, H.; Vale, P.; Blaghen, M.; Loutfi, M. Comparative study on differential accumulation of PSP toxins between cockle (*Acanthocardia tuberculatum*) and sweet clam (*Callista chione*). *Toxicon* **2005**, *46*, 612–618. [CrossRef] [PubMed]
34. European Food Safety Authority. Scientific Opinion on marine biotoxins in shellfish-Emerging toxins: Ciguatoxin group. *EFSA J.* **2010**, *8*, 1627. [CrossRef]
35. Hossen, V.; Soliño, L.; Leroy, P.; David, E.; Velge, P.; Dragacci, S.; Krys, S.; Flores Quintana, H.; Diogène, J. Contribution to the risk characterization of ciguatoxins: LOAEL estimated from eight ciguatera fish poisoning events in Guadeloupe (French West Indies). *Environ. Res.* **2015**, *143*, 100–108. [CrossRef] [PubMed]
36. Anonymous. *Domoic Acid in the King Scallop, Pecten Maximus*; Unpublished Report Prepared for the EU ASP Working Group; UK National Reference Laboratory for Marine Biotoxins: Belfast, UK, 2001.
37. Chinain, M.; Darius, H.T.; Ung, A.; Cruchet, P.; Wang, Z.; Ponton, D.; Laurent, D.; Pauillac, S. Growth and toxin production in the ciguatera-causing dinoflagellate *Gambierdiscus polynesiensis* (Dinophyceae) in culture. *Toxicon* **2010**, *56*, 739–750. [CrossRef] [PubMed]
38. Lee, R.; Lovatelli, A.; Ababouch, L. *Bivalve Depuration: Fundamental and Practical Aspects*; FAO Fisheries Technical Paper No 511; Food and Agriculture Organization of the United Nations: Rome, Italy, 2008; ISBN 9789251060063.
39. Blanco, J.; Bermúdez De La Puente, M.; Arévalo, F.; Salgado, C.; Moroño, Á. Depuration of mussels (*Mytilus galloprovincialis*) contaminated with domoic acid. *Aquat. Living Resour.* **2002**, *15*, 53–60. [CrossRef]
40. Blanco, J.; Perez-Acosta, C.; Mariño, C.; Muñiz, S.; Martín, H.; Moroño, A.; Correa, J.; Arévalo, F.; Salgado, C. Depuration of domoic acid from different body compartments of the king scallop *Pecten maximus* grown in raft culture and natural bed. *Aquat. Living Resour.* **2006**, *19*, 257–265. [CrossRef]
41. McFarland, K.; Jean, F.; Soudant, P.; Volety, A.K. Uptake and elimination of brevetoxin in the invasive green mussel, *Perna viridis*, during natural *Karenia brevis* blooms in southwest Florida. *Toxicon* **2015**, *97*, 46–52. [CrossRef] [PubMed]
42. Suzuki, T.; Ota, H.; Yamasaki, M. Direct evidence of transformation of dinophysistoxin-1 to 7-*O*-acyl-dinophysistoxin-1 (dinophysistoxin-3) in the scallop *Patinopecten yessoensis*. *Toxicon* **1999**, *37*, 187–198. [CrossRef]
43. Bauder, A.G.; Cembella, A.D.; Bricelj, V.M.; Quilliam, M.A. Uptake and fate of diarrhetic shellfish poisoning toxins from the dinoflagellate *Prorocentrum lima* in the bay scallop *Argopecten irradians*. *Mar. Ecol. Prog. Ser.* **2001**, *213*, 39–52. [CrossRef]
44. Plakas, S.M.; Wang, Z.; El Said, K.R.; Jester, E.L.E.; Granade, H.R.; Flewelling, L.; Scott, P.; Dickey, R.W. Brevetoxin metabolism and elimination in the Eastern oyster (*Crassostrea virginica*) after controlled exposures to *Karenia brevis*. *Toxicon* **2004**, *44*, 677–685. [CrossRef] [PubMed]
45. Medhioub, W.; Guéguen, M.; Lassus, P.; Bardouil, M.; Truquet, P.; Sibat, M.; Medhioub, N.; Soudant, P.; Kraiem, M.; Zouher, A. Detoxification enhancement in the gymnodimine-contaminated grooved carpet shell, *Ruditapes decussatus* (Linné). *Harmful Algae* **2010**, *9*, 200–207. [CrossRef]
46. Silvert, W.L.; Cembella, A.D. Dynamic modelling of phycotoxin kinetics in the blue mussel, *Mytilus edulis*, with implications for other marine invertebrates. *Can. J. Fish. Aquat. Sci.* **1995**, *52*, 521–531. [CrossRef]
47. Yu, K.N.; Kwong, R.W.M.; Wang, W.-X.; Lam, P.K.S. Biokinetics of paralytic shellfish toxins in the green-lipped mussel, *Perna viridis*. *Mar. Pollut. Bull.* **2007**, *54*, 1068–1071. [CrossRef] [PubMed]
48. Manger, R.L.; Leja, L.S.; Lee, S.Y.; Hungerford, J.M.; Wekell, M.M. Tetrazolium-based cell bioassay for neurotoxins active on voltage-sensitive sodium channels: Semiautomated assay for saxitoxins, brevetoxins, and ciguatoxins. *Anal. Biochem.* **1993**, *214*, 190–194. [CrossRef] [PubMed]
49. Yogi, K.; Oshiro, N.; Inafuku, Y.; Hirama, M.; Yasumoto, T. Detailed LC-MS/MS Analysis of Ciguatoxins Revealing Distinct Regional and Species Characteristics in Fish and Causative Alga from the Pacific. *Anal. Chem.* **2011**, *83*, 8886–8891. [CrossRef] [PubMed]

MDPI
St. Alban-Anlage 66
4052 Basel
Switzerland
Tel. +41 61 683 77 34
Fax +41 61 302 89 18
www.mdpi.com

Toxins Editorial Office
E-mail: toxins@mdpi.com
www.mdpi.com/journal/toxins

www.ingramcontent.com/pod-product-compliance
Lightning Source LLC
Chambersburg PA
CBHW062105220926
51850CB00010B/2905
9783039364459